Food Technology

Objective Food Microbiology

NIPA GENX ELECTRONIC RESOURCES & SOLUTIONS P. LTD.

New Delhi-110 034

Food Technology
Objective Food Microbiology

Suresh Chandra
Associate Professor
Department of Agricultural Engineering
College of Post Harvest Technology & Food Processing
Sardar Vallabhbhai Patel University of Agriculture & Technology
Meerut-250110 (U.P.) India

Ratnesh Kumar
Department of Agricultural Engineering, College of Technology
Sardar Vallabhbhai Patel University of Agriculture & Technology
Meerut-250110 (U.P.) India

Ruchi Verma
Department of Agricultural Engineering, College of Technology
Sardar Vallabhbhai Patel University of Agriculture & Technology
Meerut-250110 (U.P.) India

NIPA GENX ELECTRONIC RESOURCES & SOLUTIONS P. LTD.
New Delhi-110 034

NIPA GENX ELECTRONIC
RESOURCES & SOLUTIONS P. LTD.

101,103, Vikas Surya Plaza, CU Block
L.S.C.Market, Pitam Pura, New Delhi-110 034
Ph : +91 11 27341616, 27341717, 27341718
E-mail:newindiapublishingagency@gmail.com
www: www.nipabooks.com

For customer assistance, please contact
Phone: + 91-11-27 34 17 17 Fax: + 91-11- 27 34 16 16
E-Mail: feedbacks@nipabooks.com

ISBN: 978-93-94490-18-5

Composed and Designed by NIPA.

Dedicated to

The Love of Mother

Preface

Food Technology is the science and application of scientific, as well as socio-economic knowledge and legal rules for production. Food technology uses and exploits knowledge of Food Science and Food Engineering to produce varied foods. Study of Food Technology gives in-depth knowledge of science and technology, and develops skills for selection, storage, preservation, processing, packaging, distribution of safe, nutritious, wholesome, desirable as well as affordable, convenient foods. Another significant aspect of food technology is to promote sustainability to avoid waste and save and utilize all the food produced and ensure safe and sustainable processing practices. The demand for processed, packed and convenient food with prolonged shelf life requires well-trained human resource in the food industry. As this field requires the application of science and technology to the processing, utilization, preservation, packaging and distribution of food and food products, it encompasses a diverse range of specializations. This book covers only one sections of syllabus of GATE-Food Technology i.e. Food Microbiology.

Section II (Food Microbiology) contains five chapters which cover short notes and multiple-choice question on the syllabus as Characteristics of microorganisms, Microbial growth: growth and death kinetics, serial dilution technique. Food spoilage: spoilage microorganisms in different food products, Toxins from microbes: pathogens and non-pathogens including *Staphylococcus, Salmonella, Shigella, Escherichia, Bacillus, Clostridium,* and *Aspergillus* genera. Fermented foods and beverages.

The essential goal of this manuscript is to provide complete and simplified reach out to understanding of the basic fundamentals of Food Technology (Food Chemistry & Nutrition and Food Microbiology) to the students of the Food Technology. Although, Food Technology covers broad area which cannot confined in a single book, however our efforts have made to cover two section of Food Technology syllabus. This book is also beneficial to those students preparing who are ambitious of higher studies or going to appear in competitive examination such as GATE/NET/ARS/FSSAI examination etc. We are sure to release second book with remaining three volumes as Objective Food Technology.

Finally, we are making a plea to those who make use of this book to supply us with information or points of view that differ with those expressed in. We know that there will be errors, for which we alone are responsible, and we will appreciate the opportunity to correct those. In subsequent editions or volumes, there will be opportunity not only to correct, but to expand the perspective and thus to achieve more complete our objective of providing the serious students with a consistent and correct answer of how we make better. Feedback from readers of this book will certainly help us to improve its quality in future. The authors express thanks to their friends, students and family members who encourage for writing such book which is helpful for competitive examinations.

We know how hard it is to develop the content of a book. However, we believe that the production of a professional book of this nature is even more difficult. We thank the NIPA, New Delhi to coordinate the entire project.

March , 2022

Suresh Chandra
Ratnesh Kumar
Ruchi Verma

Contents

GATE Syllabus
XE-G Food Technology

Section-1: Food Chemistry and Nutrition: Carbohydrates: structure and functional properties of mono-, oligo-, & poly-saccharides including starch, cellulose, pectic substances and dietary fibre, gelatinization and retrogradation of starch. Proteins: classification and structure of proteins in food, biochemical changes in post mortem and tenderization of muscles. Lipids: classification and structure of lipids, rancidity, polymerization and polymorphism. Pigments: carotenoids, chlorophylls, anthocyanins, tannins and myoglobin. Food flavours: terpenes, esters, aldehydes, ketones and quinines. Enzymes: specificity, simple and inhibition kinetics, coenzymes, enzymatic and non-enzymatic browning. Nutrition: balanced diet, essential amino acids and essential fatty acids, protein efficiency ratio, water soluble and fat-soluble vitamins, role of minerals in nutrition, co-factors, anti-nutrients, nutraceuticals, nutrient deficiency diseases. Chemical and biochemical changes: changes occur in foods during different processing.

Section-2: Food Microbiology: Characteristics of microorganisms: morphology of bacteria, yeast, mold and actinomycetes, spores and vegetative cells, gram-staining. Microbial growth: growth and death kinetics, serial dilution technique. Food spoilage: spoilage microorganisms in different food products including milk, fish, meat, egg, cereals and their products. Toxins from microbes: pathogens and non-pathogens including *Staphylococcus, Salmonella, Shigella, Escherichia, Bacillus, Clostridium,* and *Aspergillus* genera. Fermented foods and beverages: curd, yoghurt, cheese, pickles, soya-sauce, sauerkraut, idli, dosa, vinegar, alcoholic beverages and sausage.

Section-3: Food Products Technology: Processing principles: thermal processing, chilling, freezing, dehydration, addition of preservatives and food additives, irradiation, fermentation, hurdle technology, intermediate moisture foods. Food packaging and storage: packaging materials, aseptic packaging, controlled and modified atmosphere storage. Cereal processing and products: milling of rice, wheat, and maize, parboiling of paddy, bread, biscuits, extruded products and ready to eat breakfast cereals. Oil processing: expelling, solvent extraction, refining and hydrogenation. Fruits and vegetables processing: extraction, clarification, concentration and packaging of fruit juice, jam, jelly, marmalade, squash, candies, tomato sauce, ketchup, and puree, potato chips, pickles. Plantation crops processing and products: tea, coffee, cocoa, spice, extraction of essential oils and oleoresins from spices. Milk and milk products processing: pasteurization and sterilization, cream, butter, ghee, ice-cream, cheese and milk powder. Processing of animal products: drying, canning, and freezing of fish and meat; production of egg powder. Waste utilization: pectin from fruit wastes, uses of by-products from rice milling. Food standards and quality maintenance: FPO, PFA, AGMARK, ISI, HACCP, food plant sanitation and cleaning in place (CIP).

Section 4: Food Engineering: Mass and energy balance; Momentum transfer: Flow rate and pressure drop relationships for Newtonian fluids flowing through pipe, Reynolds number. Heat transfer: heat transfer by conduction, convection, radiation, heat exchangers. Mass transfer: molecular diffusion and Fick's law, conduction and convective mass transfer, permeability through single and multilayer films. Mechanical operations: size reduction of solids, high pressure homogenization, filtration, centrifugation, settling, sieving, mixing & agitation of liquid. Thermal operations: thermal sterilization, evaporation of liquid foods, hot air drying of solids, spray and freeze-drying, freezing and crystallization. Mass transfer operations: psychrometry, humidification and dehumidification operations.

1

Short Notes on Food Microbiology

HISTORICAL DEVELOPMENTS

Food Preservation 1782 — Canning of vinegar was introduced by a Swedish chemist.

1810 — Preservation of food by canning was patented by Appert in France. — Peter Durand was issued a British patent to preserve food in "glass, pottery, tin or other metals or fit materials." The patent was later acquired by Hall, Gamble, and Donkin, possibly from Appert.

1813 — Donkin, Hall, and Gamble introduced the practice of post processing incubation of canned foods. — Use of SO_2 as a meat preservative is thought to have originated around this time.

1825 — T. Kensett and E. Daggett were granted a U.S. patent for preserving food in tin cans.

1835 — A patent was granted to Newton in England for making condensed milk.

1837 — Winslow was the first to can corn from the cob.

1839 — Tin cans came into wide use in the United States. L.A. Fastier was given a French patent for the use of brine bath to raise the boiling temperature of water.

1840 — Fish and fruit were first canned.

1841 — S. Goldner and J. Wertheimer were is issued British patents for brine baths based on Fastier's method.

1842 — A patent was issued to H. Benjamin in England for freezing foods by immersion in an ice and salt brine.

1843 — Sterilization by steam was first attempted by I. Winslow in Maine.

1845 — S. Elliott introduced canning to Australia.

1853 — R. Chevallier-Appert obtained a patent for sterilization of food by autoclaving.

1854 — Pasteur began wine investigations. Heating to remove undesirable organisms was introduced commercially in 1867-1868.

1855 — Grim wade in England was the first to produce powdered milk.

1856 — A patent for the manufacture of unsweetened condensed milk was granted to Gail Borden in the United States.

1861 — I. Solomon introduced the use of brine baths to the United States.

1865 — The artificial freezing offish on a commercial scale was begun in the United States. Eggs followed in 1889.

1874 — The first extensive use of ice in transporting meat at sea was begun. — Steam pressure cookers or retorts were introduced.

1878 — The first successful cargo of frozen meat went from Australia to England. The first from New Zealand to England was sent in 1882.

1880 — The pasteurization of milk was begun in Germany.

1882 — Krukowitsch was the first to note the destructive effects of ozone on spoilage bacteria.

1886 — A mechanical process of drying fruits and vegetables was carried out by an American, A.F. Spawn.

1890 — The commercial pasteurization of milk was begun in the United States. — Mechanical refrigeration for fruit storage was begun in Chicago.

1893 — The Certified Milk movement was begun by H.L. Coit in New Jersey.

1895 — The first bacteriological study of canning was made by Russell.

1907 — E. Metchnikoff and co-workers isolated and named one of the yogurt bacteria, Lactobacillus bulgaricus. — The role of acetic acid bacteria in cider production was noted by B.T.P. Barker.

1908 — Sodium benzoate was given official sanction by the United States as a preservative in certain foods.

1916 — The quick freezing of foods was achieved in Germany by R. Plank, E. Ehrenbaum, and K. Reuter.

1917 — Clarence Birdseye in the United States began work on the freezing of foods for the retail trade. — Franks was issued a patent for preserving fruits and vegetables under CO_2.

1920 — Bigelow and Esty published the first systematic study of spore heat resistance above 212°F. The "general method" for calculating thermal processes was published by Bigelow, Bohart, Richardson, and Ball; the method was simplified by CO. Ball in 1923.

1922 — Esty and Meyer established z = 18°F for Clostridium botulinum spores in phosphate buffer.

1928 — The first commercial use of controlled atmosphere storage of apples was made in Europe (first used in New York in 1940).

1929 — A patent issued in France proposed the use of high-energy radiation for the processing of foods. — Birdseye frozen foods were placed in retail markets.

1943 — B.E. Proctor in the United States was the first to employ the use of ionizing radiation to preserve hamburger meat.

1950 — The D value concept came into general use.

1954 — The antibiotic nisin was patented in England for use in certain processed cheeses to control clostridial defects.

1955 — Sorbic acid was approved for use as a food preservative. — The antibiotic chlorotetracycline was approved for use in fresh poultry (oxytetracycline followed a year later). Approval was rescinded in 1966.

1967 — The first commercial facility designed to irradiate foods was planned and designed in the United States. The second became operational in 1992 in Florida.

1988 — Nisin accorded GRAS (generally recognized as safe) status in the United States.

1990 — Irradiation of poultry approved in the United States.

1997 — The irradiation of fresh beef up to a maximum level of 4.5 kGy and frozen beef up to 7.0 kGy was approved in the United States.

1997 — Ozone was declared GRAS by the U.S. Food and Drug Administration for food use.

Food Spoilage

1659 — Kircher demonstrated the occurrence of bacteria in milk; Bondeau did the same in 1847.

1680 — Leeuwenhoek was the first to observe yeast cells.

1780 — Scheele identified lactic acid as the principal acid in sour milk.

1836 — Latour discovered the existence of yeasts.

1839 — Kircher examined slimy beet juice and found organisms that formed slime when grown in sucrose solutions.

1857 — Pasteur showed that the souring of milk was caused by the growth of organisms in it.

1866 — L. Pasteur's Etude sur Ie Vin was published.

1867 — Martin advanced the theory that cheese ripening was similar to alcoholic, lactic, and butyric fermentations.

1873 — The first reported study on the microbial deterioration of eggs was carried out by Gayon. — Lister was first to isolate Lactococcus lactis in pure culture.

1876 — Tyndall observed that bacteria in decomposing substances were always traceable to air, substances, or containers.

1878 — Cienkowski reported the first microbiological study of sugar slimes and isolated Leuconostoc mesenteroides from them.

1887 — Forster was the first to demonstrate the ability of pure cultures of bacteria to grow at 0° C.

1888 — Miquel was the first to study thermophilic bacteria.

1895 — The first records on the determination of numbers of bacteria in milk were those of Von Geuns in Amsterdam. — S.C. Prescott and W. Underwood traced the spoilage of canned corn to improper heat processing for the first time.

1902 — The term psychrophile was first used by Schmidt-Nielsen for microorganisms that grow at 0° C.

1912 — The term osmophilic was coined by Richter to describe yeasts that grow well in an environment of high osmotic pressure.

1915 — Bacillus coagulans was first isolated from coagulated milk by B.W. Hammer.

1917 — Bacillus stearothermophilus was first isolated from cream-style corn by RJ. Donk.

1933 — Oliver and Smith in England observed spoilage by Byssochlamys fulva; first described in the United States in 1964 by D. Maunder.

Food Poisoning

1820 — The German poet Justinus Kerner described "sausage poisoning" (which in all probability was botulism) and its high fatality rate.

1857 — Milk was incriminated as a transmitter of typhoid fever by W. Taylor of Penrith, England.

1870 — Francesco Selmi advanced his theory of ptomaine poisoning to explain illness contracted by eating certain foods.

1888 — Gaertner first isolated Salmonella enteritidis from meat that had caused 57 cases of food poisoning.

1894 — T. Denys was the first to associate staphylococci with food poisoning.

1896 — Van Ermengem first discovered Clostridium botulinum.

1904 — Type A strain of C. botulinum was identified by G. Landman.

1906 — Bacillus cereus food poisoning was recognized. The first case of diphyllobothriasis was recognized.

1926 — The first report of food poisoning by streptococci was made by Linden, Turner, and Thorn.

1937 — Type E strain of C. botulinum was identified by L. Bier and E. Hazen.

1937 — Paralytic shellfish poisoning was recognized.

1938 — Outbreaks of Campylobacter enteritis were traced to milk in Illinois.

1939 — Gastroenteritis caused by Yersinia enterocolitica was first recognized by Schleifstein and Coleman.

1945 — McClung was the first to prove the etiologic status of Clostridium perfringens (welchii) in food poisoning.

1951 — Vibrio parahaemolyticus was shown to be an agent of food poisoning by T. Fujino of Japan.

1955 — Similarities between cholera and Escherichia coli gastroenteritis in infants were noted by S. Thompson — Scombroid (histamine-associated) poisoning was recognized. — The first documented case of anisakiasis occurred in the United States.

1960 — Type F strain of C. botulinum identified by Moller and Scheibel — The production of aflatoxins by Aspergillus flavus was first reported.

1965 — Foodborne giardiasis was recognized.

1969 — C. perfringens enterotoxin was demonstrated by C.L. Duncan and D.H. Strong — C. botulinum type G was first isolated in Argentina by Gimenez and Ciccarelli.

1971 — First U.S. foodborne outbreak of Vibrio parahaemolyticus gastroenteritis occurred in Maryland. — First documented outbreak of E. coli foodborne gastroenteritis occurred in the United States.

1975 — Salmonella enterotoxin was demonstrated by L.R. Koupal and R.H. Deibel.

1976 — First U.S. foodborne outbreak of Yersinia enterocolitica gastroenteritis occurred in New York. — Infant botulism was first recognized in California.

1977 — The first documented outbreak of cyclosporiasis occurred in Papua, New Guinea; first in United States in 1990.

1978 — Documented foodborne outbreak of gastroenteritis caused by the Norwalk virus occurred in Australia.

1979 — Foodborne gastroenteritis caused by non-01 Vibrio cholerae occurred in Florida. Earlier outbreaks occurred in Czechoslovakia (1965) and Australia (1973).

1981 — Foodborne listeriosis outbreak was recognized in the United States.

1982 — The first outbreaks of foodborne hemorrhagic colitis occurred in the United States.

1983 — Campylobacter jejuni enterotoxin was described by Ruiz-Palacios et al.

1985 — The irradiation of pork to 0.3 to 1.0 kGy to control Trichinella spiralis was approved in the United States.

1986 — Bovine spongiform encephalopathy (BSE) was first diagnosed in cattle in the United Kingdom.

Short notes of Food Microbiology

- Acinetobacter: These Gram-negative rods show some affinity to the family Neisseriaceae. They are strict aerobes that do not reduce nitrates. Although rod-shaped cells are formed in young cultures, old cultures contain many coccoid-shaped cells. They are widely distributed in soil and water and may be found on many foods, especially refrigerated fresh products.
- Alcaligenes: Although Gram negative, these organisms sometimes stain Gram positive. They are rods and to ferment sugars but instead produce alkaline reactions, especially in litmus milk. Non pigmented, they are widely distributed in nature in decomposing matter of all types. Raw milk, poultry products, and fecal matter are common sources.
- Aeromonas: These are typically aquatic Gram-negative rods and they produce copious quantities of gas from the fermented sugars. They are producing gas and normal inhabitants of the intestines of fish, and some are fish pathogens.
- Alteromonas: These are marine and coastal water inhabitants that are found in and on seafoods; all species require seawater salinity for growth. They are Gram-negative motile rods that are strict aerobe.
- Arcobacter: They are Gram-negative curved or S-shaped rods that are quite similar to the campylobacters except they can grow at 15°C and are aerotolerant. They are found in poultry, raw milk, shellfish, and water; and in cattle and swine products. These oxidase- and catalase-positive organisms cause abortion and enteritis in some animals.

- Bacillus: These are Gram-positive spore-forming rods that are aerobes in contrast to the clostridia, which are anaerobes. Although most are mesophiles, psychrotrophs and thermophiles exist. The genus contains only two pathogens: B. anthracis (cause of anthrax) and B. cereus. Although most strains of the latter are nonpathogens, some cause foodborne gastroenteritis.
- Campylobacter: These Gram-negative, spirally curved rods were formerly classified as vibrios. They are microaerophilic to anaerobic.
- Clostridium: These anaerobic spore-forming rods are widely distributed in nature, as are their aerobic counterparts, the bacilli. The genus contains many species, some of which cause disease in humans. Mesotrophic, psychrotrophic, and thermophilic species/strains exist; their importance in the thermal canning of foods.
- Corynebacterium: This is one of the true coryneform genera of Gram-positive, rod-shaped bacteria that are sometimes involved in the spoilage of vegetable and meat products. Most are mesotrophs, although psychrotrophs are known, and one, C. diphtheriae, causes diphtheria in humans.
- Enterobacter: These enteric Gram-negative bacteria are typical of other Enterobacteriaceae relative to growth requirements, although they are not generally adapted to the gastrointestinal tract.
- Erwinia: These Gram-negative enteric rods are especially associated with plants.
- Escherichia: This is clearly the most widely studied genus of all bacteria. Those strains that cause foodborne gastroenteritis, and E. coli as an indicator of food safety.
- Lactobacillus: They are Gram-positive, catalase-negative rods that often occur in long chains. They typically occur on most, if not all, vegetables, along with some of the other lactic acid bacteria. Their occurrence in dairy products is common.
- Lactococcus: They are Gram-positive, nonmotile, and catalase-negative, spherical, or ovoid cells, that occur singly, in pairs, or as chains. They grow at 10°C but not at 45°C. Lactic acid is the predominant end-product of fermentation.
- Listeria: They are gram-positive, nonsporing rods closely related to Brochothrix.
- Micrococcus: These Gram-positive and catalase-positive cocci are inhabitants of mammalian skin and can grow in the presence of high levels of NaCl.

- Salmonella: All members of this genus of Gram-negative enteric bacteria are considered to be human pathogens. It should be noted that the salmonellae have been placed in two species with those that affect humans placed in the species Salmonella enterica.
- Vibrio: These Gram-negative straight or curved rods are members of the family Vibrionaceae.
- Molds are filamentous fungi that grow in the form of a tangled mass that spreads rapidly and may cover several inches of area in 2 to 3 days. The total of the mass or any large portion of it is referred to as mycelium. Mycelium is composed of branches or filaments referred to as hyphae.
- Bacteria, yeasts, molds, and viruses are important in food for their ability to cause food borne diseases and food spoilage and to produce food and food ingredients.
- Several species of bacteria, molds, and yeasts are considered safe or food grade, or both, and are used to produce fermented foods and food ingredients.
- Both yeasts and molds are eukaryotic, but yeasts are unicellular whereas molds are multicellular.
- Eukaryotic cells are generally much larger (20-100 μm) than prokaryotic cells (1-10 μm). Eukaryotic cells have rigid cell walls and thin plasma membranes. The cell wall does not have peptidoglycan, is rigid, and is composed of carbohydrates.
- Molds are nonmotile, filamentous, and branched. The cell wall is composed of cellulose, chitin, or both. A mold (thallus) is composed of large numbers of filaments called hyphae. An aggregate of hyphae is called mycelium. A hypha can be non-septate, septate-uninucleate, or septate-multinucleate. A hypha can be vegetative or reproductive.
- Yeasts are widely distributed in nature. The cells are oval, spherical, or elongated, about 5-30 x 2^{-10} μm in size. They are nonmotile. The cell wall contains polysaccharides (glycans), proteins, and lipids.
- Bacteria are unicellular, most ca. 0.5-1.0 x 2.0^{-10} μm in size, and have three morphological forms: spherical (cocci), rod shaped (bacilli), and curved (comma).
- On the basis of Gram-stain behaviour, bacterial cells are grouped as Gram-negative or Gram (+) positive. Gram-negative cells have a complex cell wall containing an outer membrane (OM) and a middle membrane (MM).
- Molds are important in food because they can grow even in conditions in which many bacteria cannot grow, such as low pH, low water activity (a_w), and high osmotic pressure.

- Saccharomyces cerevisiae variants are used in baking for leavening bread and in alcoholic fermentation. They also cause spoilage of food, producing alcohol and CO_2.
- Pichia: Cells are oval to cylindrical and form pellicles in beer, wine, and brine to cause spoilage. Some are also used in oriental food fermentation. Species: *Pichia membranaefaciens*.
- Rhodotorula: They are pigment-forming yeasts and can cause discoloration of foods such as meat, fish, and sauerkraut Species: *Rhodotorula glutinis*.
- Torulopsis: Cells are spherical to oval. They cause spoilage of milk because they can ferment lactose (e.g., *Torulopsis versatilis*). They also spoil fruit juice concentrates and acid foods.
- Candida: Many species spoil foods with high acid, salt, and sugar and form pellicles on the surface of liquids. Some can cause rancidity in butter and dairy products (e.g., *Candida lipolyticum*).
- Protozoan parasites: They are eukaryotic cells and some are associated with water and foodborne outbreaks. Often wild animals, live-stocks, pets, or even humans carry these parasites and serve as source. Contaminated soil and irrigation water are known causes for the contamination of fresh produce. The major organisms are Giardia, Cryptosporidium, Cyclospora, Isospora and Toxoplasma.
- Viruses are important in food for three reasons. Some are able to cause enteric disease, and thus, if present in a food, can cause foodborne diseases. Hepatitis A and Norwalk-like or Noroviruses have been implicated in foodborne outbreaks. Several other enteric viruses, such as poliovirus, adenovirus, echo virus, and Coxsackie virus, can cause foodborne diseases.
- Bacteriophages are used to transfer genetic traits in some bacterial species or strains by a process called transduction (e.g. *Escherichia coli* or *Lactococcus lactis*).

Important Bacterial Genera

S.No.	Bacterial genera	Species important in food
Gram-Negative Aerobes Bacteria		
1.	Campylobacter	Two species, *Campylobacter jejuni* and *Cam. coli*, are foodborne pathogens. Small (0.2 x 1 µm) microaerophilic helical, motile cells found in the intestinal tract of humans, animals, and birds. Mesophiles.
2.	Pseudomonas	Straight or curved (0.5 x 5 µm); aerobes; motile rods; psychrotrophs (grow at low temperatures). Found widely in the environment. This genus includes large numbers of species. Some

		important species in foods are *Pseudomonas fluorescens, Pseudomonas aerllgillosa*, and *Pseudomonas putida*. Important spoilage bacteria, can metabolize a wide variety of carbohydrates, proteins, and lipids in foods.
3.	Acetobacter	Ellipsoid to rod-shaped (0.6 x 4 μm); occur singly or in short chains; motile or nonmotile; aerobes; oxidize ethanol to acetic acid; mesophiles. Cause souring of alcoholic beverages and fruit juices and used to produce vinegar (acetic acid). Can also spoil some fruits (rot). widely distributed in plants and in places where alcohol fermentation occurs. Important species: *Acetobacter aceti*
4.	Flavobacterium	Rods with parallel sides (0.5 x 3 μm); nonmotile; colonies colored; some species psychrotrophs. Cause spoilage of milk, meat, and other protein foods. Species: *Flavobacterium aquatile.*
5.	Alcaligenes	Rods or coccobacilli (0.5 x 1 μm); motile; present in water, soil, or fecal material; mesophiles. Cause spoilage of protein-rich foods. Species: *Alcaligenes faecalis.*
6.	Psychrobacter	The genus was created in 1986 and contains one species-*Psychrobacter immobilis*. Coccobacilli (1 x 1.5 μm) and nonmotile. Can grow at 5°C or below, show optimum growth at 20°C, and unable to grow at 35°C. Found in fish meat, and poultry products.

Gram-Negative Facultative Anaerobes

7.	Citrobacter	Straight rods (1 x4 μm); single orin pairs; usually motile; mesophiles. Found in the intestinal contents of humans, animals, and birds, and in the environment. Included in the coliform group as an indicator of sanitation. Important species: *Citrobacter freundii.*
8.	Escherichia	Straight rods (l x 3 μm); motile or nonmotile; mesophiles. Found in the intestinal contents of humans, warm· blooded animals, and birds. Many strains non-pathogenic, but some strains pathogenic to humans and animals and involved in foodborne diseases. Used as an indicator of sanitation (theoretically non-pathogenic strains) in coliform and fecal coliform groups. Important species: *Escherichia coli.*
9.	Enterobacter	Straight rods (1 x 2 μm); motile; mesophiles. Found in the intestinal contents of humans, animals, birds, and in the environment. Included in the coliform group as an indicator of sanitation. Important species: *Enterobacter aerogenes.*
10.	Erwinia	Small rods (1 x 2 μm); occur in pairs or short chains; motile; facultative anaerobes; optimum growth at 30°C. Many are plant pathogens and cause spoilage of plant products. Species: *Erwinia amylovora.*
11.	Klebsiella	Medium rods (1 x 4 μm); occur singly or in pairs; motile; capsulated; mesophiles. Found in the intestinal contents of humans, animals, and birds; soil; water; and grains. Included in the coliform group as an indicator of sanitation. Important species: *Klebsiella pneumoniae.*

12.	Proteus	Straight, small rods (0.5 x 1.5 μm); highly motile; form swarm on agar media; some grow at low temperature. Occur in the intestinal contents of humans and animals and the environment. Many involved in food spoilage. Species: *Proteus vulgaris*.
13.	Salmonella	Medium rods (1 x 4 μm); usually motile; mesophiles. There are over 2000 serovars and all are regarded as human pathogens. Found in the intestinal contents of humans, animals, birds, and insects. Major cause of foodborne diseases. Species: *Salmonella enterica* ssp. enterica.
14.	Shigella	Medium rods; nonmotile; mesophiles. Found in the intestine of humans and primates. Associated with foodborne diseases. Species: *Shigella dysenteriae*.
15.	Serratia	Small rods (0.5 x 1.5 μm); motile; colonies white, pink. or red; some grow at refrigerated temperature. Occur in the environment. Cause food spoilage. Species: *Serratia liquefaciells*.
16.	Yersinia	Small rods (0.5 x 1 μm); motile or nonmotile; can grow at 1° C. Present in the intestinal contents of animals. *Yersinia enterocolitica* has been involved in foodborne disease outbreaks.
17.	Vibrio	Curved rods (0.5 x 1.0 μm); motile; mesophiles. Found in fresh water and marine environments. Some species need NaCl for growth. Several species are pathogens and have been involved in food borne disease (*Vibrio cholerae, Vibrio parahaemolyticus and Vibrio vulnificus*), whereas others can cause food spoilage (*Vibrio alginolyticus*).

Rickettsias

18.	Coxiella	Gram-negative; nonmotile; very small cells (0.2 x 0.5 μm); grow on host cells. Relatively resistant to high temperature (killed by pasteurization). *Coxiella burnetii* causes infection in cattle and has been implicated with Q fever in humans (especially on consuming unpasteurized milk).

Gram-Positive Cocci

19.	Micrococcus	Spherical cells (0.2-2 μm); occur in pairs. tetrads, or clusters; aerobes; nonmotile; some species produce yellow colonies; mesophiles, resistant to low heat. Found in mammalian skin. It can cause spoilage. Species: Micrococcus luteus.
20.	Staphylococcus	Spherical cells (0.5-1.0 μm); occur singly, in pairs, or clusters; nonmotile; mesophiles; facultative anaerobes; grow in 10% NaCl *Staphylococcus aureus* strains are frequently involved in foodborne diseases. Staphylococcus carnosus is used for processing some fermented sausages. Main habitat is skin of humans, animals, and birds.
21.	Streptococcus	Spherical or ovoid (1 μm); occur in pairs or chains; nonmotile, facultative anaerobes; mesophiles. *Streptococcus pyogenes* is pathogenic and has been implicated in foodborne diseases; present as commensals in human respiratory tract. *Streptococcus*

thermophilus is used in dairy fermentation; can be present in raw milk; can grow at 50°C.

22. Enterococcus — Spheroid cells (1.0 μm); occur in pairs or chains; nonmotile; facultative anaerobes; some strains survive low heat (pasteurization); mesophiles. Normal habitat is the intestinal contents of humans, animals, and birds, and the environment. Can establish on equipment surfaces. Used as an indicator of sanitation. Important in food spoilage. Species: *Enterococcus faecalis*.

23. Lactococcus — Ovoid elongated cells (0.5-1.0 μm); occur in pairs or short chains; non-motile; facultative anaerobes; mesophiles, but can grow at 10°C; produce lactic acid. Used to produce many bioprocessed foods, especially fermented dairy foods. Species: *Lactococcus lactis* subsp. lactis and subsp. cremoris; present in raw milk and plants and several strains produce bacteriocins, some with a relatively wide host range against Gram-positive bacteria and have potential as food bio-preservatives.

24. Leuconostoc — Spherical or lenticular cells; occur in pairs or chains; nonmotile; facultative anaerobes; heterolactic fermentators; mesophiles. but some species and strains can grow at or below 3°C. Some are used in food fermentation. Psychrotrophic strains are associated with spoilage (gas formation) of vacuum-packaged refrigerated foods. Found in plants, meat, and milk. Species: *Leuconostoc mesenteroides* subsp. mesenteroides, *Leuconostoc lactis, Leuconostoc carnosum, Leuconostoc mesenteroides* subsp. *dextranicum* produces dextran while growing in sucrose. Several strains produce bacteriocins, some with a wide spectrum against Gram-positive bacteria, and these have potential as food bio-preservatives.

25. Pediococcus — Spherical cells (1 μm); form tetrads; mostly present in pairs; nonmotile; facultative anaerobes; homolactic fermentators; mesophiles, but some can grow at 50°C; some survive pasteurization. Some species and strains are used in food fermentation. Some can cause spoilage of alcoholic beverages. Found in vegetative materials and in some food products. Species: *Pediococcus acidilactici* and *Pediococcus pentosaceus*. Several strains produce bacteriocins, some with a wide spectrum against Gram-positive bacteria, and they can be used as food bio-preservatives.

26. Sarcina — Large, spherical cells (1-2 μm), occur in packets of eight or more; nonmotile; produce acid and gas from carbohydrates; facultative anaerobes. Present in soil, plant products, and animal feces. Can be involved in spoilage of foods of plant origin. Species: *Sarcina maxima*.

Gram-Positive, Endospore-Forming Rods

27. Bacillus — Rod-shaped, straight cells; vary widely in size (small, medium, or large; 0.5-1 x 2-10 μm) and shape (thick or thin); single or in chains; motile or nonmotile; mesophiles or psychrotrophic; aerobes

or facultative anaerobes; all form endospores that are spherical or oval and large or small (one per cell), spores are highly heat resistant. Includes many species, some of which are important in foods, because they can cause foodborne disease (*Bacillus cereus*) and food spoilage, especially in canned products (*Bacillus coagulans, Bacillus stearothermophilus*). Enzymes of some species and strains are used in food bioprocessing (Bacillus subtilis). Present in soil. dust, and plant products (especially spices). Many species and strains can produce extracellular enzymes that hydrolyze carbohydrates, proteins, and lipids.

28. Clostridium — Rod-shaped cells that vary widely in size and shape; motile or nonmotile; anaerobes (some species extremely sensitive to oxygen); mesophiles or psychrotrophic; form endospores (oval or spherical) usually at one end of the cell, some species sporulate poorly, spores are heat resistant. Found in soil, marine sediments, sewage, decaying vegetation, and animal and plant products. Some are pathogens and important in food (*Clostridium botulinum, Clostridium peifringens*) and others are important in food spoilage (*Clostridium tyrobutyricum, Clostridium saccharolyticum, Clostridium laramie*). Some species are used as sources of enzymes to hydrolyze carbohydrates and proteins in food processing.

Gram-Positive, Nonsporulating Regular Rods

29. Lactobacillus — Rod-shaped cells that vary widely in shape and size, some are very long whereas others are coccobacilli, appear in single or in small and large chains; facultative anaerobes; most species are nonmotile; mesophiles (but some are psychrotrophs); can be homo- or heterolactic fermentors. Found in plant sources, milk, meat, and feces. Many are used in food bioprocessing (*Lactobacillus delbrueckii* subsp. *bulgaricus*. *Lactobacillus helveticus, Lactobacillus plantarum*) and some are used as probiotics (*Lactobacillus acidophilus*, *Lactobacillus reuteri*, *Lactobacillus casei* subsp. *casei*). Some species can grow at low temperatures in products stored at refrigerated temperature (*Lactobacillus sake, Lactobacillus curvatus*). Several strains produce bacteriocins, of which some having a wide spectrum can be used as food bio-preservatives.

30. Carnobacterium — Similar in many characteristics to lactobacilli cells; found in meat and fish; facultative anaerobes; heterofermentative; nonmotile; can grow in foods, especially in meat products; stored at refrigerated temperature. Some strains produce bacteriocins. Species: *Carnobacterium piscicola.*

31. Brochothrix — Similar in many characteristics to lactobacilli; facultative anaerobes; homofermentative; nonmotile; found in meat. It can grow in refrigerated vacuum-packaged meat and meat products. Species: *Brochothrix thermosphacta*.

32. Listeria — Short rods (0.5 x 1 μm); occur singly or in short chains; motile; facultative anaerobes; can grow at 1°C; cells killed by pasteurization. The species are widely distributed in the

		environment and have been isolated from different types of foods. Some *Listeria monocytogenes* strains are important foodborne pathogens.

Gram-Positive, Non-sporeforming Irregular Rods

33.	Corynebacterium	Slightly curved rods; some cells stain unevenly; facultative anaerobes; nonmotile; mesophiles; found in the environment, plants, and animals. Some species cause food spoilage. *Corynebacterium glutamicum* is used to produce glutamic acid.
34.	Brevibacterium	Cells can change from rod to coccoid shape; aerobes; nonmotile; mesophiles. Two species, *Brevibacterium linens* and *Brevibacterium casei*, have been implicated in the development of the aroma in several cheese varieties (surface ripened), because of the production of sulphur compounds (such as methanethiol). In other protein-rich products they can cause spoilage (in fish). They are found in different cheeses and raw milk.
35.	Propionibacterium	Pleomorphic rods (0.5 x 2 μm); can be coccoid 'bifid' or branched; present singly or in short chains; V and Y configuration and in clumps with Chinese character-like arrangement; nonmotile; anaerobes; mesophiles. Dairy propionibacteria are used in food fermentation (*Propionibacterium freudenreichii* in Swiss cheese). Produce proline and propionic acid. Found in raw milk, Swiss cheese and silage.
36.	Bifidobacterium	Rods of various shapes; present singly or in chains; arranged in V or star like shape; nonmotile; mesophiles; anaerobes. Metabolize carbohydrates to lactate and acetate. Found in colons of humans, animals, and birds. Some species are used in probiotics (*Bifidobacterium bifidum, Bifidobacterium infantis, Bifidobacterium adolescentis*).

Important Bacterial Groups in Foods

S.No.	Bacterial Groups	Their importance in foods
1.	Lactic Acid Bacteria	They are bacteria that produce relatively large quantities of lactic acid from carbohydrates. Species mainly from genera *Lactococcus, Leuconostoc, Pediocococcus, Lactobacillus*, and *Streptococcus thermophilus* are included in this group.
2.	Acetic Acid Bacteria	They are bacteria that produce acetic acid, such as *Acetobacter aceti.*
3.	Propionic Acid Bacteria	They are bacteria that produce propionic acid & are used in dairy fermentation. Species such as *Propionibacterium freudenreichii* are included in this group.
4.	Butyric Acid Bacteria	They are bacteria that produce butyric acid in relatively large amounts. Some *Clostridium* spp. such as *Clostridium butyricum* are included in this group.
5.	Proteolytic Bacteria	They are bacteria that can hydrolyse proteins because they produce extracellular proteinases. Species in genera *Micrococcus,*

		Staphylococcus, Bacillus, Clostridium, Pseudomonas, Alteromonas, Flavobacterium, Alcaligenes, some in *Enterobacteriaceae* and Brevibacterium are included in this group
6.	Lipolytic Bacteria	They are bacteria that are able to hydrolyse triglycerides because they produce extracellular lipases. Species in genera *Micrococcus, Staphylococcus, Pseudomonas, Alteromonas*, and *Flavobacterium* are included in this group.
7.	Saccharolytic Bacteria	They are bacteria that are able to hydrolyse complex carbohydrates. Species in genera *Bacillus, Clostridium, Aeromonas, Pseudomonas*, and *Enterobacter* are included in this group.
8.	Thermophilic Bacteria	They are bacteria that are able to grow at 50^0C and above. Species from genera *Bacillus, Clostridiun, Pediococcus, Streptococcus*, and *Lactobacillus* are included in this group.
9.	Psychrotrophic Bacteria	They are bacteria that are able to grow at refrigerated temperature (d"5^0C). Some species from *Pseudomonas, Alteromonas, Alcaligenes, Flavobacterium, Serratia, Bacillus, Clostridium, Lactobacillus, Leuconostoc, Carnobacterium, Brochothrix, Listeria, Yersinia,* and *Aeromonas* are included in this group
10.	Thermoduric Bacteria	They are bacteria that are able to survive pasteurization temperature treatment. Some species from *Micrococcus, Enterococcus, Lactobacillus, Pediococcus, Bacillus* (spores), and *Clostridium* (spores) are included in this group.
11.	Halotolerant Bacteria	They are bacteria that are able to survive high salt concentrations ($\geq$10%). Some species from *Bacillus, Micrococcus, Staphylococcus, Pediococcus, Vibrio, and Corynebacterium* are included in this group.
12.	Aciduric Bacteria	They are bacteria that are able to survive at low pH ($<$ 4.0). Some species from *Lactobacillus, Pediococcus, Lactococcus, Enterococcus*, and *Streptococcus* are included in this group.
13.	Osmophilic Bacteria	They are bacteria that can grow at a relatively higher osmotic environment than that is needed for other bacteria. Some species from genera *Staphylococcus, Leuconostoc*, and *Lactobacillus* are included in this group. They are much less osmophilic than yeasts and molds.
14.	Gas-Producing Bacteria	They are bacteria that produce gas (CO_2, H_2, H_2S) during metabolism of nutrients. Species from genera *Leuconostoc. Lactobacillus, Propionibacterium, Escherichia. Enterobacter, Clostridium* and *Desulfotomaculum* are included in this group.
15.	Slime Producers	They are bacteria that produce slime because they synthesize polysaccharides. Some species or strains from *Xanthomonas, Leuconostoc, Alcaligenes, Enterobacter, Lactococcus* and *Lactobacillus* are included in this group.
16.	Spore Formers	They are bacteria having the ability to produce spores. Species from *Bacillus, Clostridium* and *Desulfotomaculum* are included in this group. They are further divided into aerobic spore-formers,

		anaerobic spore-formers), flat sour sporeformers, thermophilic sporeformers, and sulfide-producing sporeformers.
17.	Aerobes	They are bacteria that require oxygen for growth and multiplication. Species from *Pseudomonas, Bacillus*, and *Flavobacterium* are included in this group.
18.	Anaerobes	They are bacteria that cannot grow in the presence of oxygen. Species from *Clostridium* are included in this group.
19.	Facultative Anaerobes	They are bacteria that are able to grow in both the presence and absence of oxygen. *Lactobacillus, Pediococcus, Leuconostoc,* enteric pathogens. and some species of *Bacillus, Serratia,* and *coliforms* are included in this group.
20.	Coliforms	Species from *Escherichia, Enterobacter, Citrobacter* and *Klebsiella* are included in this group. They are used as an index of sanitation.
21.	Fecal Coliforms	Mainly *Escherichia coli* is included in this group. They are also used as an index of sanitation.
22.	Enteric Pathogens	Pathogenic *Salmonella, Shigella. Campylobacter, Yersinia, Escherichia. Vibrio*, hepatitis A, and others that can cause gastrointestinal infection are included in this group.

- Generally, dry air with low dust content and higher temperature has a low microbial level.
- Microbial contamination of food from the air can be reduced by removing the potential sources, controlling dust particles in the air (using filtered air), using positive air pressure, reducing humidity level, and installing UV light.
- Soil contaminated with fecal materials can be the source of enteric pathogenic bacteria and viruses in food. Different types of parasites can also get in food from soil. Removal of soil (and sediments) by washing and avoiding soil contamination can reduce microorganisms in foods from this source.
- Sewage, especially when used as fertilizer in crops, can contaminate food with microorganisms, the most significant of which are different enteropathogenic bacteria and viruses. To reduce incidence of microbial contamination of foods from sewage, it is better not to use sewage as fertilizer. If used, it should be efficiently treated to kill the pathogens. Also, effective washing of foods following harvesting is important.
- Water is used to produce, process, and, under certain conditions, store foods. It is used for irrigation of crops, drinking by food animals and birds, raising fishery and marine products, washing foods, processing (pasteurization, canning, and cooling of heated foods) and storage of foods (e.g., fish on ice), washing and sanitation of equipment, and processing and transportation facilities. Water is also used as an ingredient in many processed foods. Thus, water quality can greatly influence microbial quality of foods.

Contamination of foods with pathogenic bacteria, viruses, and parasites from water has been recorded.

- Waste water can be recycled for irrigation. However, chlorine-treated potable water (drinking water) should be used in processing, washing, sanitation, and as an ingredient. Although potable water does not contain coliforms and pathogens (mainly enteric types). It can contain other bacteria capable of causing food spoilage, such as Pseudomonas, Alcaligenes, and Flavobacterium. Improperly treated water can contain pathogenic and spoilage microorganisms. To overcome the problems, many food processors use water, especially as an ingredient, that has a higher microbial quality than that of potable water.
- Improperly cleaned hands, lack of aesthetic sense and personal hygiene, and dirty clothes and hair can be major sources of microbial contamination in foods. The presence of minor cuts and infection in hands and face and mild generalized diseases (e.g., flu, strep throat~ or hepatitis A in an early stage) can amplify the situation. In addition to spoilage bacteria, pathogens such as *Sta. aureus, Salmonella serovars, Shigella* spp., *pathogenic E. coli*, Norovirus, and hepatitis A can be introduced into foods from human sources sometimes through fecal-oral contamination. Proper training of personnel in personal hygiene, regular checking of health, and maintaining efficient sanitary and aesthetic standards are necessary to reduce microbial contamination from this source.
- Various spices generally have very high populations of mold and bacterial spores. Starch, sugar and flour might have spores of thermophilic bacteria. Pathogens have been isolated from dried coconut, egg, and chocolate. The ingredients should be produced under sanitary conditions and given antimicrobial treatments. In addition, setting up acceptable microbial specifications for the ingredients will be important in reducing microorganisms in food from this source.
- Many types of microorganisms from air, raw foods, water, and personnel can get into the equipment and contaminate foods. Depending on the environment (moisture, nutrients, and temperature) and time, microorganisms can multiply and, even from a low initial population reach a high level and contaminate large volumes of foods. Proper cleaning and sanitation of equipment at prescribed intervals are important to reduce microbial levels in food. In addition, developing means to prevent or reduce contamination from air, water, personnel, and insects is important.
- Microorganisms present in raw and processed (non-sterile and commercially sterile) foods are important for their involvement in foodborne diseases, food spoilage, and food bioprocessing.

- Growth or cell multiplication of bacteria, yeasts, and molds is influenced by the intrinsic and extrinsic environments of the food.
- Microbial growth in laboratory media is also important for quantitative and qualitative detection of the microbiological quality of a food.
- Bacteria reproduce by a process called transverse binary fission, or, simply, binary fission.
- A bacterial cell has a specific surface area-to-volume (s/v) ratio. A newly divided cell has a higher s/v ratio, which helps in the rapid transport of nutrients from the environment.
- Yeasts and molds can also reproduce asexually. A yeast cell produces a bud that initially is much smaller in size and remains attached to the surface of the original cell.
- Molds can grow in size by cell division or elongation at the tip of a hyphae.
- The time that a single cell takes to divide into two is called generation time. However, in practice, generation time is referred to as the doubling time for the entire population.
- Generally, in food systems, microorganisms have longer generation times than in a nutritionally rich bacteriological broth.
- The rate of growth of microorganisms during exponential growth phase can also be determined mathematically by measuring cell numbers, OD_{600} nm, cell mass (wet or dry weight), or cell constituents (proteins, RNA, or DNA).
- A fast-growing bacterial strain under optimum conditions can have a specified growth rate (μ) in unit time (h^{-1}) of 2.5 or higher.
- The specific growth rate equations are important for determining predictable growth rate and population level (or of other components) in fermentation and shelf life of foods.
- Many environmental parameters of food, such as storage temperature, acidity (pH), water activity (a_w), oxidation-reduction (O-R) potential, and nutrients, influenced the microbial growth rate.
- The growth rate and growth characteristics of a microbial population under a given condition can be graphically represented by counting cell numbers, enumerating CFUs, or measuring optical density in a spectrophotometer at a given wavelength (above 300 nm, usually at 600 nm) of a cell suspension.
- The cells in the population differ initially in metabolic rate and only some multiply, and then almost all cells multiply. This is the exponential phase (also called logarithmic phase). Growth rate at the exponential phase follows first order reaction kinetics and can be used to determine generation time.

Following this, the growth rate slows down and finally the population enters the stationary phase. At this stage, because of nutrient shortage and accumulation of waste products, a few cells die and a few cells multiply, keeping the living population stable.

- Intrinsic factors of a food include nutrients, growth factors, and inhibitors (or antimicrobials), water activity, pH, and oxidation-reduction potential.
- Water is not considered a nutrient, but it is essential as a medium for the biochemical reactions necessary for the synthesis of cell mass and energy in microbial growth.

Carbohydrates in Foods: Major carbohydrates present in different foods, either naturally or added as ingredients, can be grouped on the basis of chemical nature as follows:

Monosaccharides	• Hexoses: glucose, fructose, mannose, galactose
	• Pentoses: xylose, arabinose, ribose, ribulose, xylulose
Disaccharides	• Lactose (galactose + glucose)
	• Sucrose (fructose + glucose)
	• Maltose (glucose + glucose)
Oligosaccharides	• Raffinose (glucose + fructose + galactose)
	• Stachyose (glucose + fructose + galactose + galactose)
Polysaccharides	• Starch (glucose units)
	• Glycogen (glucose units)
	• Cellulose (glucose units)
	• Inulin (fructose units)
	• Hemicellulose (xylose, galactose, mannose units)
	• Dextrans (α-1, 6 glucose polymer)
	• Pectins
	• Gums and mucilages

- Lactose is found only in milk and thus can be present in foods made from or with milk and milk products. Glycogen is present in animal tissues, especially in liver. Pentoses, most oligosaccharides and polysaccharides are naturally present in foods of plant origin.
- Molds are the most capable of using polysaccharides.
- Food carbohydrates are metabolized by microorganisms principally to supply energy through several metabolic pathways.

- Microorganisms also produce metabolic by-products associated with food spoilage (CO_2 to cause gas defect) or food bioprocessing (lactic acid in fermented foods).
- Microorganisms can also polymerize some monosaccharides to produce complex carbohydrates such as dextrans, capsular materials, and cell wall (or outer membrane and middle membrane in Gram-negative bacteria).
- Proteins and peptides are polymers of different amino acids without or with other organic (e.g., a carbohydrate) or inorganic (e.g., iron) components and contain 15-18% nitrogen.
- Simple food proteins are polymers of amino acids, such as albumins (in egg), globulins (in milk), glutelins (gluten in cereal), prolamins (zein in grains), and albuminoids (collagen in muscle).
- Lipids are relatively higher in foods of animal origin than in foods of plant origin, although nuts, oil seeds, coconuts, and olives have high amounts of lipids.
- Cholesterols are present in foods of animal origin or foods containing ingredients from animal sources.
- Many microorganisms can produce extracellular lipases that can hydrolyse glycerides to fatty acids and glycerol.
- Water activity (a_w) is a measure of the availability of water for biological functions and relates to water present in a food in free form.
- Bound water is the fraction used to hydrate hydrophilic molecules and to dissolve solutes, and is not available for biological functions; thus, it does not contribute to a_w.
- The a_w of a food can be expressed by the ratio of water vapor pressure of the food (P, which is 0 to < 1) to that of pure water (P_0, which is 1), that is, P_0/P. It ranges between 0 and 1, or more accurately >0 to < 1, because no food can have a water activity of either 0 or 1. The a_w of a food can be determined from its equilibrium relative humidity (ERH) by dividing ERH by 100 (because ERH is expressed in percentage).
- The a_w of food ranges from 0.1 to 0.99. The a_w values of some food groups are as follows: (i) cereals, crackers, sugar, salt, dry milk, 0.10-0.20; (ii) noodles, honey, chocolate, dried egg, <0.60; (iii) jam. jelly, dried fruits, parmesan cheese, nuts, 0.60-0.85; (iv) fermented sausage, dry cured meat, sweetened condensed milk, maple syrup, 0.85-0.93; (v) evaporated milk, tomato paste, bread, fruit juices, salted fish, sausage, processed cheese, 0.93-0.98; and (vi) fresh meat, fish, fruits, vegetables, milk, eggs, 0.98-0.99.

- The a_w of foods can be reduced by removing water (desorption) and increased by the adsorption of water, and these two parameters can be used to draw a sorption isotherm graph for a food.
- The desorption process gives relatively lower Aw values than the adsorption process does at the same moisture content of a food. This has important implications in the control of a microorganism by reducing the Aw of a food.
- The a_w of a food can be reduced by several means, such as adding solutes, ions, hydrophilic colloids, and freezing and drying.
- The free water in a food is necessary for microbial growth. It is necessary to transport nutrients and remove waste materials, carry out enzymatic reactions, synthesize cellular materials, and take part in other biochemical reactions, such as hydrolysis of a polymer to monomers (proteins to amino acids).
- In general, the minimum a_w values for growth of microbial groups are as follows: most molds, 0.8, with xerophilic molds as low as 0.6; most yeasts, 0.85, with osmophilic yeasts, 0.6-0.7; most Gram-positive bacteria, 0.90; and Gram-negative bacteria, 0.93.
- On the basis of pH, foods can be grouped as high-acid foods (pH below 4.6) and low-acid foods (pH 4.6 and above).
- Most fruits, fruit juices, fermented foods (from fruits, vegetables, meat, and milk), and salad dressings are high-acid (low-pH) foods, whereas most vegetables, meat, fish, milk, and soups are low-acid (high-pH) foods.
- Tomato, however, is a high-acid vegetable (pH 4.1-4.4).
- In general, molds and yeasts are able to grow at lower pH than do bacteria, and Gram-negative bacteria are more sensitive to low pH than are Gram positive bacteria.
- The pH range of growth for molds is 1.5-9.0; for yeasts, 2.0-8.5; for Gram-positive bacteria, 4.0-8.5; and for Gram-negative bacteria, 4.5-9.0.
- Extrinsic factors important in microbial growth in a food include the environmental conditions in which it is stored. These are temperature, relative humidity and gaseous environment. The relative humidity and gaseous condition of storage, respectively, influence the a_w and E_h of the food.
- Microbial growth is accomplished through enzymatic reactions. It is well-known that within a certain range, with every 10°C rise in temperature, the catalytic rate of an enzyme doubles. Similarly, the enzymatic reaction rate is reduced to half by decreasing the temperature by 10°C.
- Psychrotrophs are microorganisms that grow at refrigerated temperature (0-5°C), irrespective of their optimum range of growth temperature. They

usually grow rapidly between 10 and 30°C. Molds; yeasts; many Gram-negative bacteria from genera *Pseudomonas, Achromobacter, Yersinia, Serratia,* and *Aeromonas*; and Gram-positive bacteria from genera *Leuconostoc, Lactobacillus, Bacillus, Clostridium*) and *Listeria* are included in this group.

- Microorganisms that survive pasteurization temperature are designated as thermodurics. They include species from genera *Micrococcus, Bacillus, Clostridium, Lactobacillus, Pediococcus*, and *Enterococcus*. Bacterial spores are also included in this group.
- The most important genus and species used is *Saccharomyces cerevisiae*. It has been used to leaven bread and produce beer, wine, distilled liquors, and industrial alcohol; produce invertase (enzyme); and flavor some foods (soups).
- *Aspergillus oryzae* is used in fermentation of several oriental foods, such as sake, soy sauce, and miso. It is also used as a source of some food enzymes.
- *Aspergillus* niger is used to produce citric acid and gluconic acid from sucrose. It is also used as a source of the enzymes pectinase and amylase.
- *Penicillium roquefortii* is used for ripening of Roquefort, Gorgonzola, and blue cheeses. Some strains can produce the neurotoxin roquefortin.
- Penicillium camembertii is used in Camembert cheese and *Penicillium caseicolum* is used in Brie cheese. They are also used to produce the enzyme glucose oxidase.
- Buttermilk: Made with Lactococcus species without or with *Leuconostoc cremoris*; some can have biovar diacetylactis in place of *Leuconostoc cremoris* (such as ymer in Denmark), whereas some can have a ropy variant of *Lactococcus* species (langfil in Norway) or mold (*Geotrichum candidum* in villi in Finland).
- Yogurt: Made with *Streptococcus thermophilus* and *Lactobacillus delbrueckii* subsp. bulgaricus; some types can also have added *Lactobacillus acidophilus*, *Lactobacillus casei*, *Lactobacillus rhamnosus*, and *Bifidobacterium* spp.; some may also have *Lactococcus* species and *Lactobacillus* plantarum and lactose-fermenting yeasts (dahi in India).
- Acidophilus Milk: Made with Lactobacillus acidophilus.
- Bifidus milk: Made with *Bifidobacterium* spp.
- Yakult. Made with *Lactobacillus casei*; may contain *Bifidobacterium* spp.
- Kefir: Made from *Lactobacillus* kefir (several species of yeasts along with *Leuconostoc, Lactobacillus*, and *Lactococcus* spp.).

- Kumiss: Made from Lactobacillus delbrueckii subsp. *bulgaricus* and yeasts.
- Sauerkraut fermentation: It is produced by fermenting shredded cabbage. The product has a sour taste with a clean acid flavor. The raw material has a large number of undesirable organisms and a small population of lactic acid bacteria < 1 %). Among the lactic acid bacteria, most are Lactococcus spp. and Leuconostoc spp., and a small fraction is Lactobacillus spp. and Pediococcus spp. The presence of 2.25 % salt, large amounts of fermentable sugars (sucrose, hexoses, pentoses), absence of oxygen, and low fermentation temperature facilitate Leuconostoc spp., primarily Leuconostoc mesenteroides, to grow rapidly in sauerkraut. The characteristic flavor of sauerkraut is the result of the combined effects of lactate, acetate, ethanol, CO_2, and diacetyl in proper amounts.
- Sausage fermentation: Semidry sausages include summer sausage, thuringer, and semidry salami. The average composition is 30% fat, 20% protein. 3% minerals (salts), and 47% water. They have a tangy taste with a desirable flavor imparted by the combined effect of lactate, acetate, and diacetyl, and some breakdown components from proteolysis and lipolysis. The use of spices also contributes to the flavor. Those containing nitrite have a pinkish color in contrast to the grayish color in products without it. For high temperature and low pH, Pediococcus acidilactici strains are preferred; for low temperature and high pH, Lab. plantarum strains are preferred. Pediococcus pentosaceus strains can be used under both conditions. Some starters can have both Pediococcus and Lactobacillus species. In addition, selected Micrococcus spp. or Sta. carnosus strains are added as secondary flora for their beneficial effects on desired product color. The optimum growth temperatures for Pediococcus acidilactici, Pediococcus pelltosaceus, and Lab. plantarum are 40, 35, and 30°C (l04, 95, and 86°F) respectively. Micrococcus spp. and Sta. carnosus grown at 32.2°C (90°F). Cooking to an internal temperature of 60°C (140°F) kills Latobacillius plantarum and probably Pediococcus pentosaceus, but probably not Pediococcus acidilactici, Micrococcus, or Sta. Carnosus.
- Cultured buttermilk fermentation: Cultured buttermilk is produced from partially skim milk through controlled fermentation with starter cultures. It processed with following steps i.e. (i) Skim milk: 9% SNF + citrate (0.2%); (ii) Heated at 185°F (85°C) for 30 min (kills bacterial cells and phages), (iii) Cooled to 72°F (22°C), starter added, agitated for 50 min (incorporates air), (iv) Incubated at 72°F (22°C) for 12 h, pH 4.7, acidity 0.9%; (v) Gel broken, cooled to 400 P (4.5°C), and salted (package). In this Lac. lactis spp. lactis or cremoris is used for acid and Leu. mesenteroides spp. cremoris for diacetyl and CO_2.

- Yogurt Fermentation: Plain yogurt has a semisolid mass due to coagulation of milk (skim, low or full fat) by starter-culture bacteria. It has a sharp acid taste with a flavor similar to walnuts and a smooth mouth feel. The flavor is due to the combined effects of acetaldehyde, lactate, diacetyl and acetate, but 90% of the flavor is due to acetaldehyde. Many types of yogurt are available in the market for example, plain yogurt, fruit yogurt, Havored and colored yogurt, blended yogurt, sweetened yogurt, heated yogurt, frozen yogurt, dried yogurt, low-lactose yogurt, and carbonated yogurt. Frozen concentrates or direct vat set starters can be used. Normally, Lab delbrueckii spp. bulgaricus and Str. thermophilus are used. For a good product, the two starter species should be added at a Streptococcus : Lactobacillus cell ratio of 1:1; in the final product, the ratio should not exceed 3:2. The major flavor compound in yogurt is acetaldehyde (25 ppm), with some diacetyl (0.5 ppm) and acetate. Acetaldehyde is produced in two ways: from glucose via pyruvate by Streptococcus sp. and from threonine (supplied or produced through proteolysis in milk) by *Lactobacillus* sp.

- In plain yogurt, flavor problems can be associated with the concentrations of acetaldehyde. A low concentration gives a chalky and sour flavor, and too much acetaldehyde can give a green flavor. Similarly, too much diacetyl gives a buttery aroma. Too much acid production during storage causes a sour taste. Proteolysis and accumulation of bitter peptides during storage are associated with bitter flavor. Production of exopolysaccharides by the starter can give a viscous and ropy texture (which can be desirable in some situations). Growth of yeasts during storage can also produce a fruity flavof, especially in yogurt containing fruits and nuts. In colored, flavored and blended yogurt, many of these problems are masked.

- Cottage cheese is made from low-fat or skim milk and has a soft texture with ca. 80% moisture. It is unripened and has a buttery aroma due to diacetyl (along with lactic acid and little acetaldehyde). Mixed strains of Lac. lactis spp. cremoris and lactis are predominantly used for acid. Leu. mesenteroides spp. cremoris can be added initially (in which case citrate is added to milk), mainly for diacetyl, Lac. lactis spp. lactis biovar diacetylactis can be used for diacetyl, but not inoculated in milk because of formation of too much CO_2 that causes curd particles to float. Processing of cottage cheese (from Skim Milk) i.e. **1.** Pasteurized, cooled to 70°F (22.2°C), starter added, and incubated for 12 h at pH 4.7; **2.** Firm curd set, cut in cubes, and cooked at 125°F (51.7°C) for 50 min or more; **3.** Whey drained off, stirred to remove more to get dry curd; **4.** Salted, creamed, and preservative added; **5.** Packaged and refrigerated.

Cheeses fermentation: Cheeses are made by coagulating the casein in milk with lactic acid produced by lactic acid bacteria, without or with the enzyme rennin, followed by collecting the casein for further processing, which may include ripening.

Unripened Cheese	Soft	• Cottage cheese with starters Lac. lactis spp. lactis and cremoris and Leuconostoc mesenteroides ssp. cremoris.· Mozzarella cheese with starters Stl: thermophilus and Lab. delbrueckii ssp. bulgaricus.
Ripened Cheese	Soft	• Brie cheese with starter Lac. lactis spp.; Penicillium spp. and yeasts are secondary flora.
	Semihard	• Gouda cheese with starters Lac. lactis ssp. and *Leuconostoc* spp.; dairy Propionibacterium may be secondary flora. Blue cheese with starter Lac. lactis ssp., Leuconostoc spp. Penicillium roquefortii, yeasts, and micrococci are secondary flora.
	Hard	• Cheddar cheese with starters Lac. lactis spp.; some lactobacilli and pediococci (and probably enterococci) are secondary (or associative) flora.
		• Swiss cheese with starters Str. thermophilus. Lab. helveticus, and dairy *Propionibacterium* spp.; enterococci can be secondary (or associative) flora.

- Cheddar cheese is made from whole milk contains less than 39% moisture. 48% fat, is generally orange-yellow in color (due to added color, annatto) and ripened. Smoothness of texture and intensity of characteristic flavor vary with the starters used and the period of ripening. The typical flavor is the result of a delicate balance among flavor components produced during ripening through enzymatic breakdown of carbohydrates, proteins, and lipids in unripened cheddar cheese. Starters are selected mixed strains of Lac. lactis spp. cremoris or lactis. Leuconostoc, which may be added for flavor. Starters can be used as frozen concentrates (direct vat set). Processing of Cheddar cheese i.e. 1. Pasteurized; color (annatto) and starter added; 2. Incubated at 86°P (30°C) for acidity to increase by 0.2% and rennet added; 3. Incubated for coagulation (30 min) and cut in cubes; 4. Cooked at 100°F (37.8°C), whey drained, cheddaring to lose whey and curd to mat; 5. Milled, salted, put in form, pressed for 16 h to drain whey, and removed; 6. Dried for 5 days at 50^0 F (10°C), waxed, vacuum packaged, and cured at 40^{0}F (4.4^{0}C) for 2-12 months.
- Swiss cheese is made from partially skimmed milk (cow's milk) and coagulated with acid and rennin; it is hard and contains 41 % moisture and 43% fat The cheese should have uniformly distributed medium-sized eyes

(openings). It has a sweet taste, due to proline, and a nutty flavor. Starters are Str. thermophilus and Lab. helveticus as primary for acid and Propionibacterium sp. as secondary for eye formation, taste, and flavor. Its Processing is as 1. Pasteurized, starter added, and incubated at 90⁰F (32.2°C) for acidity to increase by 0.2%; 2. Rennin added, incubated for firm coagulation, cut in 1/8-in. cubes, and cooked for 1 h at 125°F (51.7°C); 3. whey removed, curd pressed for 16 h, cut in blocks, and exposed in brine at 55°F (12.8°C) for 1-3 days; 4. Surface dried, vacuum packaged, stored for 7 days at 55°F (12.8°C), and transferred to 75°F (23.9°C) for 1-4 weeks; 5. Cured at 37°F (2.8°C) for 3-9 months..

- In Swiss cheese, Spores of Clostridium tyrobutyricum, present in raw milk or entering as contaminants, can germinate. grow, and cause rancidity and gas blowing in this low-acid cheese. Nisin has been used as a bio-preservative to control.
- Blue cheese is a semihard (46% moisture, 50% fat), mold-ripened cheese made from whole milk (cow's milk). It has a crumbly body, mottled blue color and sharp lipolytic flavor. Lac. lactis spp. cremoris or lactis and Leu.. cremoris or lactis serve as primary starters. Pen. roqueform spores serve as secondary starters. The lactic starters grow until curing, and from lactose, they produce lactate, diacetyl, acetate, CO_2) and acetaldehyde. Mold spores, during storage at 50°F (10°C) in high humidity, germinate quickly, produce mycelia, and spread inside to give the mottled green appearance. Their growth continues during curing. Puncturing the inside of the cheese helps remove CO_2 and let's air in to help the growth of molds. Its Processing is as 1. Homogenized, pasteurized, starter added and incubated at 90⁰F (32.2°C) for acidity to increase to 0.2%; 2. Rennet added, incubated for firm set, cut, cooked at 100⁰F (37.8°C), and whey drained; 3. Curd collected in hoop. drained 16 h, and salted in brine for 7 days. 4. Spiked to let air get inside. Maid spores added stored at 500 P (10°C) in high humidity for 4 weeks. 5. Stored at 40°F (4.4°C) for curing for 3 months.
- Soy sauce (Shoyu) is used as a seasoning to increase the appetite and to add a delicious, spicy flavor and color (38). It is a light brown to black liquid with a meat like, salty flavor produced by hydrolyzing soybeans, with or without the addition of wheat or barley, using enzymes produced by Aspergillus oryzae (A. soyae) and the action of lactic acid bacteria and yeast in salt water of high concentration.
- Soy sauces are salty liquid seasonings ranging from amber to dull brown color, produced from soy beans/defatted soy grits with or without wheat, barley, and/or rice by a two-stage fermentation, an aerobic solid-state fungal fermentation is followed by an anaerobic mixed lactic-yeast, submerged

fermentation in a strong salt brine. The industrial production of the Japanese-type soy sauce (koikuchi shoyu) can be divided into the following stages (i) Treatment of raw materials. (ii) Koji production. (iii) Moromi preparation and fermentation (aging). (iv) Pressing of aged mash. (v) Pasteurization and refining.

- Good quality shoyu contains 1.5-1.8% (g=100 mL) total nitrogen, of which 40-50% is lower peptides and peptones, and 40-50% amino acids, of which approximately 20% is glutamic acid. It contains 2-5% reducing sugar, 60% of which is glucose; 1-2% v/v alcohol, 1-2% organic acids (60-80% of which is lactic acid); 18% v/v (15% w/w) sodium chloride; and a pH of 4.6-4.9.
- Fresh meats from food animals and birds contain a large group of potential spoilage bacteria that include species of Pseudomonas, Acinetobacter, Moraxella, Shewanella, Alcaligenes, Aeromonas. Escherichia, Enterobacter, Serratia, Hafnia, Proteus, Brochothrix, Micrococcus, Enterococcus~ Lactobacillus, Leuconostoc, Carnobacterium, and Clostridium, as well as yeasts and molds.
- To delay microbial spoilage, fresh meats are stored at refrigerated temperature (-5°C), unless the facilities are not available. Thus, normally psychrotrophic bacteria are the most predominant types in raw meat spoilage.
- Refrigerated meat in a modified atmosphere, such as in a mixture of CO_2 and O_2, favors growth of facultative anaerobic Brochothrix thermosphacta, especially in meat with pH 6.0 or higher (DFD meat).
- To reduce spoilage of fresh meats, initial microbial level should be reduced. In addition, storage at low temperatures (close to 0 to -1 °C), modified atmosphere packaging, and vacuum packaging should be done.
- Evaporated milk is condensed whole milk with 7.5% milk fat and 25% total solids. It is packaged in hermetically sealed cans and heated to obtain commercial sterility. Under proper processing conditions, only thermophilic spores of spoilage bacteria can survive, and exposure to high storage temperature (43°C or higher) can trigger their germination and subsequent growth. Under such conditions, Bacillus species, such as Bac. coagulans, can cause coagulation of milk (flakes, clots, or a solid curd).
- Condensed milk is generally condensed whole milk and has 10-12% fat and 36% total solids. The milk is initially given a low-heat treatment, close to pasteurization temperature and then subjected to evaporation under partial vacuum (at 50°C). Thus, it can have thermoduric microorganisms that subsequently can grow and cause spoilage. Other microorganisms can also

get into the product during the condensing process. Even at refrigerated temperature, this product has a limited shelf life, as does pasteurized milk.

- Sweetened condensed milk contains 8.5% fat, 28% total solids, and 42% sucrose. The whole milk is initially heated to a high temperature (80-100°C) and then condensed at 60°C under vacuum and put into containers. Because of a low A_w, it is susceptible to spoilage from the growth of osmophilic yeasts (such as *Torula* spp.), causing gas formation. If the containers have enough headspace and oxygen, molds (e.g., *Penicillium* and *Aspergillus*) can grow on the surface.
- Butter contains 80% milk fat and can be salted or unsalted. The microbiological quality of butter depends on the quality of cream and the sanitary conditions used in the processing. Growth of bacteria (*Pseudomonas* spp.), yeasts (*Candida* spp.), and molds (Geotrichum candidum) on the surface causes flavor defects (putrid, rancid, or fishy) and surface discoloration. In unsalted butter, coliforms, Enterococcus, and Pseudomonas can grow favourably in the water phase (which has nutrients from milk) and produce flavour defects.
- Grains and seeds normally have 10-12% moisture, which lowers the a_w to d" 0.6 and thus inhibits microbial growth.
- In cereal grains, during harvesting, processing, and storage, if the a_w increases above 0.6, some molds can grow. Some species of storage fungi from genera Aspergillus, Penicillium, and Rhizopus can cause spoilage of high-moisture grains.
- Refrigerated dough (for biscuits, rolls, and pizzas) is susceptible to spoilage (gas formation) from the growth of psychrotrophic heterolactic acid bacterial species of Lactobacillus and Leuconostoc. Rapid CO_2 production can blow the containers, especially when the storage temperature increases to 10°C and above.
- The A_w of breads is normally low enough (0.75-0.9) to prevent growth of bacteria. However, some molds (bread molds: Rhizopus stolonifer) can grow, especially if moisture is released because of starch crystallization during storage.
- Liquid sweeteners include honey, sugar syrups, maple syrups, corn syrups, and molasses. Confectionery products include soft-centered fondant, cream, jellies, chocolate, and Turkish delight. Most of these products have an a_w of 0.8 or below, and are normally not susceptible to bacterial spoilage. Under aerobic conditions, some xerophilic molds can produce visible spoilage. However, osmophilic yeasts from genera Zygosaccharomyces (Zygosaccharomyces rollxii), S. accharomyce (Saccharomyces cerevisiae),

Torulopsis (Torulopsis holmu), and Candida (Candida valida) can ferment these products. To prevent growth of yeasts in some of these products with slightly higher a_w (such as in maple syrup), chemical preservatives are added.

- Canned foods are heat treated to kill microorganisms present, and the extent of heat treatment is predominantly dependent on the pH of a food. However, spores of some spoilage bacteria, which have greater heat resistance than do spores of *Clo. botulinum*, can survive. Thus, these products are called commercially sterile (instead of sterile, which means free of any living organism) foods.

- Canned food spoilage is both due to non microbial (chemical and enzymatic reactions) and microbial reasons. Production of hydrogen (hydrogen swell), CO_2, browning, corrosion of cans due to chemical reactions, and liquification, gelation, and discoloration due to enzymatic reactions are some examples of non microbial spoilage. Microbial spoilage is due to three main reasons: (1) inadequate cooling after heating or high-temperature storage, allowing germination and growth of thermophilic sporeformers; (2) inadequate heating, resulting in survival and growth of mesophilic microorganisms (vegetative cells and spores); and (3) leakage (can be microscopic) in the cans, allowing microbial contamination from outside following heat treatment and their growth.

- Thermophilic sporeformers: Thermophilic sporeformers can cause three types of spoilage of low-acid (high-pH) foods (such as corn, beans, and peas) when the cans are temperature abused at 43°C and above, even for a short duration.

- Flat Sour Spoilage: The cans do not swell but the products become acidic because of germination and growth of facultative anaerobic Bac. stearothermaphilus. Germination occurs at high temperature (43°C and above), but growth can take place at lower temperature (30°C and above). The organism ferments carbohydrates to produce acids without gas but with some off-flavor and cloudiness.

- Thermophilic Anaerobe (TA) Spoilage: The spoilage is caused by the growth of anaerobic Clo. thermosaccharolyticum with the production of large quantities of H_2 and CO_2 gas and swelling of cans, with sour and cheesy odor. Following germination in the thermophilic range (43°C and above), the cells can grow at lower temperatures (30°C and above).

- Sulfide Stinker Spoilage: This spoilage is caused by the Gram-negative anaerobic sporeformer Desulfotomaculum nigrificans. The spoilage is characterized by a flat container but darkened products with the odor of rotten eggs due to H_2S produced by the bacterium. H_2S, produced from the sulfur-containing amino acids, dissolves in the liquid and reacts with iron to form black color of iron sulfide. Both germination and growth occur at thermophilic range (43°C and above).

- Intoxication: Illness occurs as a consequence of ingestion of a preformed bacterial or mold toxin (mycotoxin) due to its growth in a food. A toxin has to be present in the contaminated food in an active form. Once the microorganisms have grown and produced toxin in a food, there is no need of viable cells during the consumption of the food for illness to occur. Example: Staphylococcal food poisoning.
- Infection: Illness occurs as a result of the consumption of food and water contaminated with enteropathogenic bacteria or viruses. It is necessary that the cells of enteropathogenic bacteria and viruses remain alive in the food or water during consumption. The viable cells, even if present in small numbers, have the potential to establish and multiply in the digestive tract to cause the illness (e.g., Salmonellosis, hepatitis A).
- Toxico-infection: Illness occurs from the ingestion of a large number of viable cells of some pathogenic bacteria through contaminated food and water. Generally, the bacterial cells sporulate, colonize, or die. and release toxin(s) to produce the symptoms (e.g., Clo. perfringens gastroenteritis).
- Staphylococcal intoxication: Staphylococcal food poisoning (also known as-staphylococcal gastroenteritis; staph food poisoning), caused by toxins of Staphylococcus aureus, is considered to be one of the most frequently occurring foodborne diseases worldwide. Sta. aureus are Gram-positive cocci, occur generally in grape-like clusters and are non motile, non capsular, and non sporulating. The cells are killed at 66°C in 12 min, and at 72°C in 15s. Sta. aureus are facultative anaerobes, but grow rapidly under aerobic conditions. They can ferment carbohydrates and also cause proteolysis by the extracellular proteolytic enzymes. Enterotoxigenic strains of Sta. aureus produce 17 different enterotoxins: A, B, Cl, C2, C3, D, and E through R (also designated as SEA, SEB, etc.). SEB is more stable than SEA and it has been considered as a potential weapon in bioterrorism. Normal temperature and time used in processing or cooking foods will not destroy the potency of the toxins.
- Botulism: Foodborne botulism results following consumption of food containing the potent botulinum toxin of Clostridium botulinum. It is a neurotoxin and produces neurological symptoms along with some gastric symptoms. Infant botulism occurs from the ingestion by the infant of Clo. botulinum spores that germinate, grow, and produce toxins in the GI tract and cause specific symptoms. Cells of Clo. botulinum strains are Gram-positive rods and occur

as single cells or in small chains; many are motile, obligate anaerobes and form single terminal spores. Cells are sensitive to low pH <4.6), low a_w (0.93), and moderately high salt (5.5%). Spores do not germinate in the presence of nitrite (250 ppm). Spores are highly heat resistant (killed at 115°C) but the cells are killed at moderate heat (pasteurization). Toxins form during growth. Clo. botulinum strains, on the basis of the type of toxin produced, have been divided into six types: A, B, C, D and F of these A, B, E and F have been associated with human food borne intoxications. Type A strains are proteolytic, type E strains are non proteolytic, but types Band F strains can be either proteolytic or non proteolytic. The botulinum neurontoxin (BoNT) is a 150 kDa protein toxin produced by Clo. botulinum. It is an A-B type toxin consisting of two subunits: A subunit is 50 kDa and B is 100 kDa.

- Mycotoxicosis: Many strains of molds, while growing in a suitable environment (including in foods), produce metabolites that are toxic to humans, animals, and birds, and are grouped as mycotoxins. 17-20 consumption of foods containing mycotoxins causes mycotoxicosis. Some of the toxigenic strains from several species and genera and the toxins they produce include Asp. flavus, Asp. parasiticus (both produce aflatoxins), Asp. nidulans and Asp. virsicolor (sterigmatocystin), Fusarium verticilioides (fumonisin). Fus. graminearum (deoxynivalenol-DON) Fusarium spp. (Zearalenone), Penicillium viridicatllm (ochratoxin), Pen. patulum (patulin), Pen. roqueforti (roquefortin) and Claviceps purpurea (ergot alkaloids).

Microbial Foodborne Diseases and Causative Pathogens

Types of disease	Causative microorganism	Major symptom type
Intoxication		
Staph poisoning	Staphylococcus aureus strains	Vomiting, diarrhea
Botulism	Clostridium botulinum strains	Neurologic
Mycotoxin poisoning	Mycotoxins producing strains (e.g., Aspergillus flavus)	Carcinogenic, Hepatotoxic
Infection		
Salmonellosis	Over 2000 Salmonella enterica serovars (except Sal. Typhi and Paratyphi)	Diarrhea
Campylobacter enteritis	Campylobacter jejuni and Cam. coli strains	Diarrhea
Yersiniosis	Pathogenic strains of Yersinia enterocolitica	Diarrhea
Enterohemorrhagic Escherichia coli (EHEC)	Esc. coli O157:H7, Esc. coli 026:H11	Haemorrhagic diarrhea, Hemolytic uremic syndrome (HUS)
Enteropathogenic Esc. coli (EPEC)	Esc. coli 0111:H12	Hemorrhagic diarrhea
Shigellosis	Four Shigella species (e.g., Shi, dysenterae)	Bloody mucoid dirrehea
Vibrio parahaemolyticus gastroenteritis	Pathogenic strains of Vib. parahemolyticus	Diarrhea, hepatitis
Vibrio vulnificus infection	Vib. vulnificus strains	Diarrhea, hepatitis
Brucellosis	Brucella abortus	Gastric and nongastric
Listeriosis	Listeria monocytogenes	Fever, meningitis, abortion, diarrhea (rare)
Viral infections	Pathogenic enteric viruses (e.g., Hepatitis A virus)	Fever, diarrhea, hepatitis
Madcow disease	Bovine spongiform encephalopathy (BSE) (e.g. prion)	Neurologic
Toxicoinfection		
Clostridium perfringens gastroenteritis	Clo, perfringens	Diarrhea, vomiting
Bacillus cereus gastroenteritis	Bacillus cereus strains	Vomiting, diarrhea
Escherichia coli gastroenteritis	Enterotoxigenic Esc. coli (ETEC) serotype 015:H11	Travellers diarrhea
Cholera	Vibrio cholerae	Diarrhea
Gastroenteritis by opportunistic pathogens		
Aeromonas hydrophila Gastroenteritis	Aeromonas hydrophila	Diarrhea
Plesiomonas shigelloide gastroenteritis	Plesiomonas shigelloides	Diarrhea

- A bacterial food intoxication therefore refers to food-borne illnesses caused by the presence of a bacterial toxin formed in the food.
- A bacterial food infection refers to food borne illnesses caused by the entrance of bacteria into the body through ingestion of contaminated foods and the reaction of the body to their presence or to their metabolites.
- Staphylococcus Food Intoxication: The food poisonings are caused by the ingestion of the enterotoxin formed in food during growth of certain strains of Staphylococcus aureus. The toxin is termed an enterotoxin because it causes gastro- enteritis or inflammation of the lining of the intestinal tract. S. aureus produces six serologically distinct enterotoxins (A, B, C_1, C_2, D, and E) that differ in toxicity; most food poisoning is from type A. A million staphylococci per milliliter or gram of perishable foods will be inactivated by 66 C maintained for at least 12 min or by 60 C at 78h to 83 min.
- The Enterotoxin: The staphylococcal enterotoxins are simple proteins with molecular weights between 26,000 and 30,000. The single polypeptide chains are cross-linked by a disulfide bridge to form a characteristic cystine loop. Toxin is produced at an appreciable rate at temperatures between 15.6 and 46.1 C, and production is best at 40 C. About 75 percent of all staphylococcal food-poisoning outbreaks occur because of inadequate cooling of foods. The most common human symptoms are salivation, then nausea, vomiting, retching, abdominal cramping of varying severity, and diarrhea. Blood and mucus may be found in stools and vomitus in severe cases. Headache, muscular cramping, sweating, chills, prostration, weak pulse, shock, and shallow respiration may occur. The means of prevention of outbreaks of staphylococcus food poisoning include (1) prevention of contamination of the food with the staphylococci, (2) prevention of the growth of the staphylococci, and (3) killing staphylococci in foods.
- Salmonellosis may result following the ingestion of viable cells of a member of the genus Salmonella. The Salmonella infections that are called food poisoning may be caused by any of a large number of serovars. The term serovar is used to distinguish strains of different antigenic complements. The salmonellae are gram-negative non-spore-forming rods that ferment glucose, usually with gas, but usually do not ferment lactose or sucrose. Like other bacteria, they will grow over a wider range of temperature, pH, and a_w in a good culture medium rather than in a poor one. Recommendations for thermal destruction of salmonellae in perishable foods are similar to those for staphylococci, namely, heating to 66 C and holding all parts at that temperature for at least 12 min (or 78 to 83 min at 60 C). F_{140} values (minutes at 140 F necessary to reduce an inoculum to an undetectable level) found for two species were 78 and 19 min, respectively, in custard, and 81.5

and 3.1 min in chicken á la king. These results illustrate how the required heat treatment differs with the species of Salmonella and the food heated. The principal symptoms of a Salmonella gastrointestinal infection are nausea, vomiting, abdominal pain, and diarrhea that usually appear suddenly. This may be preceded by a headache and chills.

- Clostridium perfringens Gastroenteritis: The bacterium causing this illness is C. perfringens (welchii), type A, a grampositive, nonmotile, anaerobic, spore-forming rod. Maximal temperature for growth is about 55 C, and optimal temperature is about 43 to 47 C. Growth is restricted at 15 to 20 C. The presence of a large number of C. perfringens in a cooked food is indicative of mishandling.
- Vibrio Parahaemolyticus Infection and Other Vibrios: V. parahaemolyticus is a gram-negative straight or curved motile rod. It is a halo-phile (requiring 1 to 3 percent NaCl) and will grow in 7.0 percent NaCl. Optimal temperature for growth is 35 to 37 C, but it will grow over a range of 10 to 44 C. Outbreaks of V. parahaemolyticus gastroenteritis have received considerable attention in Japan, where it is one of the most commonly occurring food-poisoning syndromes. Food-borne outbreaks of Vibrio cholerae and non-ol strains have been associated with the consumption of oysters. Vibrio vulnificus can also be isolated from seafoods and seawater. The organism is highly invasive, releasing both a hemolysin and a cytotoxin, and can result in primary septicemia in humans.
- Escherichia Coli: E. coli is generally regarded as part of the normal flora of the human intestinal tract and that of many animals. Serotypes of E. coli which have been implicated in human diarrheal diseases or food poisoning outbreaks have been designated enteropathogenic E. coli (EEC).
- Shigellosis: Food-borne outbreaks of shigellosis have been reported in the United States, but most incidents of shigellosis involve contaminated water. Optimal temperature for growth is 37 C, with a temperature range of 10 to 40 C. The organisms tolerate salt concentrations of 5 to 6 percent and are relatively heat sensitive. Pathogenicity involves the release of a lipopolysaccharide endotoxin which affects the intestinal mucosa.
- Yersinia Enterocolitica: Yersinia enterocolitica is a small rod-shaped bacteria that can cause gastrointestinal illness in humans. It has been isolated from the intestinal tracts and feces of many animals, including cats, pigs, dogs, deer, raccoons, and horses. The pig appears to be the main reservoir for the strains causing infection in humans. Y. enterocolitica has been isolated from many foods, including beef, pork, liquid eggs, soft cheese, raw milk, pasteurized milk. fish, raw oysters. shrimps, crabs, chocolate milk, turkey,

chowmein, powdered milk, and tofu. The usual symptoms, including severe abdominal pain, fever, and diarrhea, occur 24 to 36 hr after consumption of the product.

- Campylobacter: The heat-tolerant Campylobacter, i.e., C. jejuni and C. coli, are the strains most frequently associated with acute gastroenteritis in humans. The term "heat- tolerant" is a misnomer since it refers to growth at 42°C and not to any degree of heat tolerance or heat resistance; actually, these strains are inactivated at temperatures above 45 to 50°C. Their most distinguishing characteristics are absence of growth below 25°C, sensitivity to air (21 percent O_2), resistance to cephalothin, a generation or doubling time of about 1 hr at optimum temperature (42°C), sensitivity to drying, sensitivity to acid conditions, and being a poor competitor in mixed populations because of an inability to utilize carbohydrates. Symptoms include abdominal pain, cramps, diarrhea, headache, fever, and occasionally bloody stools. The incubation time is usually 2 to 3 days but may be as long as 7 to 10 days.
- Plesiomonas Shigelloides: Plesiomonas shigelloides is a facultatively anaerobic gram-negative rod. Most strains are motile by polar flagella and can grow on minimal media containing ammonium salts as the nitrogen source and glucose as the carbon source. The organism can be pathogenic in humans, causing disturbances of the intestinal tract. Diarrhea in various degrees of severity is the major symptom.
- Listeria Monocytogenes: Listeria monocytogens is a gram-positive, motile, short rod capable of growth at 4°C. L. monocytogenes has been isolated from water, milk, silage, sewage, and the feces of many animals, including humans.
- The microorganisms that cause food spoilage are moulds, yeasts and Bacteria.
- Aflatoxins produced by moulds growing on ground nut.
- All enzymes are inactivated by temperature above 80°C.
- Antifungal food additives are basically chemicals that prevent the mold growth.
- Lye solution is made from caustic soda.
- Moromi is partial product of soy sauce.
- Acidophilus milk is heated at 98^{0}C temperature for 30 min.
- Milk of 0.5-1.8 percent fat used for preparation of culture butter milk.
- Kefir is foamy and fizzy in nature due to carbon dioxide.
- The disease caused by the toxins of *Clostridium botulinum* is known as botulism.

- Mycotoxins produced by moulds such as *Aspergillus flavous* and *A. parasiticus* are aflatoxin.
- The illness caused by the consumption by bacterial toxin formed in the foods refers as food intoxication.
- The illness caused by the entrance of bacteria into the body through ingestion of contaminated food is called food infection.
- Lathyrism, a disease is associated with the consumption of khesari dhal.
- Soaking, heating or fermentation of pulses can be eliminated most of the antinutrtional factors.
- Idli is the most popular breakfast of South India.
- The preservative effect of heat processing is due to denaturation of protein in microorganisms.
- The time needed to destroy 90% of the microorganisms (to reduce their number by a factor of 10) is referred to as decimal reduction time or D-value.
- Spores are much more heat resistance than vegetable cells.
- The destruction of many vitamins, aroma compounds and pigments by heat follows similar first order reactions to microbial destruction.
- Water in foods exerts a vapour pressure.
- The ratio of the vapour pressure of water in food to the saturated vapour pressure of water at the same temperature is defined as water activity.
- Almost all microbial activity in food is inhibited below water activity (a_w) of 0.60.
- The interaction of water activity with temperature, pH, oxygen and carbon dioxide or chemical preservatives has an important effect on the inhibition of microbial growth.
- When different values of relative humidity versus equilibrium moisture content are plotted, a curve known as water absorption isotherm.
- Naturally produces peptides that inhibit other microorganisms, similar in effect to antibiotics is referred to as bacteriocins.
- An exotoxin produces by Cl. botulinum, able to cause 'fatal' food poisoning is called botulin.
- The identification of potentially hazardous ingredients, storage conditions, packaging, critical process points and relevant human factors which may affect product safety or quality is called hazard analysis.

- A package that is designed to be secure against entry of microorganisms and maintain the commercial sterility of its contents after processing is known as hermetically sealed container.
- The space in a bottle between the surface of a food and the underside of the lid is called head space.
- Microorganisms that produce more than one metabolic product are called hetero-fermentative.
- Microorganisms that produce single metabolic product are called homo-fermentative.
- The reduction in size and increase in number of solid or liquid particles in the dispersed phase is called homogenization.
- Integrated effect of heating temperature and time on microorganisms is called lethality.
- Green leafy vegetables are rich in vitamins and minerals.
- Microorganisms are more sensitive to rapid heating than chemical reaction.
- Spiral in shape is possessed by the spirilla and vibrios.
- Microorganisms will grow at as high as 82^{0}C is known as thermophilic.
- The organisms can grow under either aerobic or anaerobic conditions is called facultative.
- Bacteria are multiply by cell division process.
- Under favourable condition, bacteria can double their number every 30 minute.
- Aflatoxin in foods is produced by *Aspergillus flavous*.
- When food is handled at 10-38^{0}C, the rate of chemical reactions is approximately doubled for every 10^{0}C rise in temperature.
- Thermophiles will grow in the range of 66-82^{0}C.
- Most of bacteria are killed in range of 82-93^{0}C.
- Bacteria and yeasts generally require more moisture than molds for growth and multiplication.
- Water activity is the ratio of vapour pressure of the solution to the vapour pressure of pure water at the same temperature.
- Water activity is equal to ERH/100.
- Acid combined with heat makes the heat more destructive to microorganisms.
- Solutions of high solute concentration have a high osmotic pressure and a low water activity.

- Dilute solutions are low in osmotic pressure and a low water activity.
- Yeasts and molds are more tolerant than most bacteria against osmosis, sugar and salt.
- Cottage cheese may have sorbic acid mixed directly in the product.
- Food irradiation is also known as cold sterilization.
- The toxin that can be formed by clostridium botulinum is destroyed by a 10 min exposure to moist heat at 100^{0}C.
- Most canned or bottled food products are commercially sterile and have a shelf life of 2 years or more.
- The time in minutes at a specified temperature required to destroy 90% of the organisms in a population is known as F_0 value.
- The Z-value is the number of degree required for a specified thermal death time curve to pass through one log cycle (change by a factor of 10).
- F-value is defined as the number of minutes at a specified temperature required to destroy a specified number of organism having a specified Z-value.
- The F_0 – value is also known as the 'sterilization value'.
- The point in can or mass of food which is last to reach the final heating temperature is known as 'cold point'.
- Hydrogen peroxide is used in aseptic processing and packaging to sterilize the packaging materials.
- Rapid sterilization at extremely high temperature is referred as UHT sterilization.
- Tomato juice under HTST heating is sterilized at 121^{0}C for 0.7 min.
- In microwave heating of foods, heat transfer will not from outside to inside.
- Commercial and household refrigerators are usually operated at 4.5 to 7.0^{0}C.
- At constant temperature, a decrease in pressure increases the rate of boiling.
- Although a large proportion of the microbial load is killed during drying operations but many bacterial spore remains to survive.
- Thawing is a reverse process of freezing.
- Most of the fish are less stable in nature.
- Hydrogen peroxide is a strong oxidizing agent and a biological poison.
- Irradiation can be inhibited sprouting of potatoes and onions by specific doses.

- Sauerkraut is not an unfermented food which made from cabbage.
- Miso is a fermented food product of soybean.
- Industrial production of riboflavin and B_{12} is due to special fermentation process.
- Lactose (milk sugar) fermented by *Streptococcus lactic* bacteria is supported to yield cottage cheese.
- In the production of sauerkraut, approximately 2.0-2.5% salt generally is added to the cabbage.
- Fermented meat sausages commonly have acidities in the range of pH of 4.0-4.5.
- Ageing or ripening of meat generally is done at 2°C.
- Curing refers to modification of meat that affects preservation, flavour, colour and tenderness due to added curing ingredients.
- Sodium nitrate or nitrite is helped to cure meat and have anti-botulinum activity.
- Soy sauce is a dark brown liquid with a salty taste made by soybean and roasted wheat.
- Strains of *Saccharomyces rouxii, torulopsis* yeasts *and Pediococcus soyai* were found to be important flavour producers in soy sauce.
- Acidophilus milk is produced by fermentation of milk by *Lactobacillus acidophilus.*
- Koumiss is similar to kefir except that mare's milk is used in its preparation.
- Mare's milk is low in casein content and does not curdle like cow's milk.
- Cultured butter milk is produced from pasteurized skim milk or part skim milk cultured with lactic and aroma producing organism. *Streptoccus lactis* and/or *Streptoccus cremoris*, *S. lactis subsp. diacelylactis* and *Leuconostoc cremoris* are frequent cultures.
- Proteins in cheese are break down by rennin combined with protease to poly peptides and amino acids which contribute to the flavour but make the casein softer.
- Milk fat (lipids) in cheese production is broken down by enzymes into the glycerol and fatty acids in presence of the lipase.
- Sauerkraut is made from cabbage by natural fermentation in presence of LAB and 2 to 3% salt.
- The relative humidity which is in equilibrium with the grain moisture varies with the temperature.

2

Food Microbiology-I

1. Who discovered bacteria?

   a) Leeuwen hock b) Ledge berg

   c) Walkman and Zinder d) None

2. Smallest cell/organs are that of

   a) Vibrio b) Bacillus

   c) Mycoplasma d) Rhizobium

3. Bacteria which retain purple color after staining with Gram stain is

   a) Gram+ve b) Gram–ve

   c) Trichous d) Spirulina

4. The common mode of reproduction in bacteria is

   a) Fission b) Budding

   c) Sexual reproduction d) Sporulation

5. Typhoid is caused by

   a) Xanthomonas b) Bacillus desenteriae

   c) Salmonella typhi d) Bacillus diplococcic

6. Bacteria can't survive in highly salted pickle because

   a) Salts inhibit reproduction

   b) Bacteria do not get enough light for photosynthesis

   c) They become plasmolyzed and consequently killed

   d) The pickle does not contain nutrients necessary for bacteria to live

7. Pasteurization means

   a) Vaccination for a baby against small pox

   b) Sterilization in steam cooker at 100^0C for 10 min

   c) Heating milk or other liquid to 60^0C to 70^0C for short duration

   d) A technique or curing people bitten by mad dogs

8. Bacteria differ from other plants in that they don't have
   a) DNA b) RNA
   c) Cell wall d) A well-defined nucleus
9. Bacteria having a tuft of flagella at one end are called
   a) Peritrichorus b) Monotrichorus
   c) Lophotrichorus d) Amphitrichorus
10. Bacteria were regarded to be plants because
   a) Some of them are green b) They are present every where
   c) Some of them cannot more d) They have a rigid cell wall
11. Vinegar is prepared from fermented sugar solution by the activities of
   a) Acetobacter aceti b) Bacillus aceti
   c) B. Subtilis d) Diplococcus
12. Streptococcus lactic is responsible of
   a) Conversion of molasses into alcohol
   b) Conversion of milk into curd
   c) Tanning of leather
   d) Flavoring the leaves of tea and tobacco
13. Cholera is caused by
   a) Bacillus mycobacterium b) Vibrio cholera
   c) Pseudomonas citri d) Streptococcus cholera
14. The preparation & flavoring of leaver of tea & tobacco is due to activities of
   a) Streptococcus lactis b) Bacillus megatherium
   c) Acetobacter d) Bacillus redicicola
15. The bacteria genome is called
   a) Nucleus b) Nucleolus
   c) Nucleoid d) None
16. Escherichia coli is a bacterium which is common inhabitant of
   a) Human intestine b) Soil
   c) Milk d) Water
17. The cell of cyanobacteria and bacteria exhibit similarity in having
   a) Plastids b) Nuclei
   c) Centro some d) Naked DNA

18. Clostridium butulinum has been used in the synthesis of
    a) Vitamin B b) Vitamin A
    c) Vitamin C d) Vitamin D
19. Antibiotics are mostly obtained from
    a) Bacteria b) Viruses
    c) Angiosperms d) Fungi
20. Different types of flavor of tea and tobacco are largely due to the
    a) Mechanization process b) Activity of fungi
    c) Activity of certain bacteria d) Activity of viruses
21. Bacteria are classified on the basis of
    a) Nucleus b) Cell wall
    c) Gram positive and Gram negative d) Method of nutrition
22. A compound which is produced by organism and inhabits the growth of other organism is called
    a) Antiseptic b) Anticoagulant
    c) Antibiotic d) Antiallergic
23. Bacteria resemble
    a) Nostoc species b) Mitochondria
    c) Chlamydomonas d) None
24. Bacteria have incipient nucleus (nucleoid) and hence they are placed in
    a) Prokaryota b) Eukaryota
    c) Fungi d) Protista
25. Bacteria responsible for fermentation of dairy milk is
    a) Lactobacillus b) Hay bacillus
    c) Acetobacter d) Rhizobium
26. We can keep food for longer duration in cold storage than in ordinary cupboard because
    a) Insects cannot cause infection
    b) Bacterial multiplication is completely percentile
    c) Bacterial multiplication is greatly reduced
    d) Low temperature causes plasmolysis

27. Bacterial bearing flagella all over body are called
    a) Peritrichous
    b) Atricious
    c) Monotrichous
    d) Cephalotrichous
28. Botulism is a
    a) Type of food poisoning due to saprophytic bacterium
    b) Disease in man due to parasitic bacterium
    c) Disease in various organisms
    d) Disease of plants due to virus
29. Plasmids
    a) Virus
    b) New types of microorganisms
    c) Extra chromosomal genetic element of bacteria
    d) Genetic element of bacteria
30. Streptomycin is produced by
    a) Streptomycin griseus
    b) Streptomyces fradie
    c) Streptomyccs Venezuelae
    d) Sterptomyces scoleus
31. All bacteria have the following organelle
    a) Mesosomes
    b) Golgi bodies
    c) Mitochondria
    d) Chloroplast
32. Rickettsiae is a group of
    a) Viruses
    b) Microorganisms
    c) Bacteria
    d) PPLO
33. Vinegar is produced through
    a) Fermentation of sugar by lactobacillus
    b) Fermentation of sugar by Aspergillums
    c) Fermentation of sugar by saccharomyces
    d) A two step of process, first involving fermentation of sugar by yeast & second involving fermentation of ethyl alcohol by acetic acid bacteria
34. The disease caused by bacteria is
    a) Amoebic dysentery
    b) Beriberi
    c) Arthritis
    d) Diphtheria
35. Some bacteria have capsule outside cell wall. It is made of
    a) Proteins
    b) Cellulose
    c) Fat
    d) Mucopolysaccharide

36. Cell membranes certain branched chain lipid in
    a) Actinomycetes
    b) Spiro chaetes
    c) Eubacteria
    d) Archaebacteria
37. Tetanus is a disease caused by
    a) Virus
    b) Bacterium
    c) Fungus
    d) Inset
38. Which of the following is not true of Escherichia coli?
    a) Gene recombination can occur through transformation, transduction and conjugation
    b) It occurs in human intestine
    c) It lacks true nuclear
    d) It is diploid
39. Bacteria reproduction is
    a) Only asexual
    b) Only sexual
    c) Mostly asexual
    d) Mostly sexual
40. Which state is correct?
    a) All bacteria are autotrophic
    b) All bacteria are heterotrophic
    c) All bacteria are photosynthetic
    d) Mostly bacteria are heterotrophic but some are autotrophic
41. Bacteria are thought to be primitive organisms because they
    a) Are small, microscopic plant, which cannot be seen by naked eye
    b) Cause serious diseases in human beings, domesticated animals and crop plants
    c) Produce endospores which are very resistant to adverse condition
    d) Possess incipient nucleus and show division similar to amitosis
42. Which not a bacterial action?
    a) N_2 fixation
    b) Emulsification of fat
    c) Sewage disposal
    d) Ripening of cream
43. Cyanobacteria are
    a) Saprotrophs
    b) Photoautrophs
    c) Chemoautotrophic
    d) Chemohetrotrophic

44. Gram positive bacteria differ from gram negative bacteria in the structure of their

   a) Nucleoid / genophore b) Cytoplasm
   c) Cell wall d) Ribosomes

45. Terramycin is obtained from

   a) Streptomyces rimosus b) Streptomyces griseus
   c) Streptomyces renezuelae d) Streptomyces aureofacience

46. A spiral bacterium is included in

   a) Coccus b) Bacillus
   c) Diplococcus d) None

47. In unfavorable conditions bacteria produce resting spores called

   a) Exospores b) Chlamydospores
   c) Oidia d) Endospores

48. Food poisoning is caused by

   a) Clostridium botulinum b) Salmonella typhosa
   c) Clostridium tetani d) None of these

49. The role of bacteria in retting of fibers is hydrolysis of

   a) Cellulose of the cell walls of the fibres
   b) Lignin of the secondary wall
   c) Living content of the cells
   d) Pectic substances that bind the cells together

50. Bacteria which can also live in the absence of oxygen are

   a) Obligate aerobes b) Facultative aerobes
   c) Obligate anaerobes d) Facultative anaerobes

51. The flagella of bacteria are composed of

   a) Carbohydrate b) Lipid
   c) Protein d) Amide

52. Bacteria and yeast are similar in all the following features expect that

   a) Both are unicellular b) Both are prokaryotes
   c) Both are capable of causing fermentation d) Both produce spores

53. The similarity between Cyanobacterium and bacterium is
   a) Presence of flagella b) Presence of 80s ribosome
   c) Presence of nucleoid d) None of these
54. The genetic material of bacteria is known as
   a) Gene b) Chromosome
   c) Genophore d) Nucleohistone
55. The organisms which are included in the kingdom Monera are
   a) Unicellular b) Without a definite nucleus
   c) Uninucleate d) Coenocytes
56. A cell wall material presents only in blue green algal and bacteria is
   a) Muramic acid b) Cellulose
   c) Chitin d) Pectin
57. Blue color of blue green algal is due to
   a) Phycocyanin and allophycocyanin b) Phycoerythrin
   c) Anthocyanin d) Anthoxanthin
58. Clostridium botulinum is
   a) Obligate aerobe b) Facultative aerobe
   c) Facultative anaerobes d) Obligate anaerobes
59. One of the following is less resistance or more susceptible to antibiotics
   a) Gram +ve bacteria b) Gram –ve bacteria
   c) Escherichia coli d) None of these
60. Putrefying bacteria act upon
   a) Fats b) Carbohydrates
   c) Protein d) Starch
61. Grape like aggregates of coccus bacteria constitute
   a) Sarcina b) Staphylococcus
   c) Streptococcus d) Diplococcus
62. Pasteurization is performed at
   a) 100^0C for 15 min b) 82^0C for 30 mi
   c) 72^0C for 20 min d) 62^0C for 30 min

63. Bacterial activity is minimum at
    a) Low temperature
    b) High temperature
    c) Moderate temperature
    d) Moderately high temperature
64. Comma-shaped bacteria are
    a) Vibrio
    b) Spirillum
    c) Spirochaete
    d) Coccus
65. Clostridium is an example of
    a) Obligate aerobic N_2 fixing bacteria
    b) Facultative aerobic N_2 fixing bacteria
    c) Non-nitrogen fixing bacteria
    d) Aerobic nitrogen fixing bacteria
66. The cell wall of bacteria contains
    a) Cellulose
    b) Hemicellulose
    c) Peptidoglycan
    d) All the above
67. Surface appendages or hairy structure present in some bacteria for attaching to another and the hosts are
    a) Pili
    b) Flagella
    c) Cilia
    d) Mesosomes
68. Rod shaped bacteria are
    a) Mycobacteria
    b) Cocci
    c) Vibrios
    d) Bacilli
69. Which is wrong about bacteria
    a) Their cell wall contain peptidoglycan
    b) They were the first living beings
    c) They have prokaryotic structure
    d) Bacteria multiply by mitosis
70. A biodegradable plastic can be obtained from
    a) Rhodococcus equi
    b) Pseudomonas species
    c) Ochrobacterium species
    d) All the above
71. Anton van Leeuwenhock, the first to discover bacteria is from
    a) U.K.
    b) Sweden
    c) France
    d) Holland

72. When milk is heated at 62^0C for 30 min and then cooled, the process in called

a) Sterilization
b) Pasteurization
c) Nitrification
d) Freezing

73. Pasteurization frees food stuffs from

a) All bacteria
b) All living organisms
c) Vegetative forms of bacteria
d) Vegetative forms of all pathogenic bacteria

74. Photosynthetic bacteria include

a) Nitrobacteria and nitrosamonas
b) Chlorobium and Rhodospirillum
c) Streptococcus and staphylococcus
d) Chlorobium and clostridium

75. Bacteria without flagella are

a) Atrichous
b) Lophotrichous
c) Amphitrichous
d) Peritrichous

76. Wine turns sour

a) On exposure to light
b) Contamination by aerobic bacteria Acetobacter aceti
c) Contamination by anaerobic bacteria
d) On heating

77. The technique of developing pure culture of bacteria was developed by

a) J. Lister
b) R. Koch
c) L. Pasteur
d) A.V. Leeuwen hock

78. Rounded bacteria are

a) Bacilli
b) Vibrio
c) Spirilla
d) Cocci

79. Food poisoning and gas forming rod-shaped bacterium is

a) Shigella
b) Salmonella
c) Clostridium
d) E. Coli

80. Bacteria generally moves due to

a) Chemotaxis
b) Thermotaxis
c) Phototaxis
d) Thermotropism

81. A peculiar amino acid present in bacterial cell wall is
    a) Glutamate b) Alanine
    c) Diaminopimelic acid d) Asparate
82. Which is rod shaped?
    a) Vibrio cholarea b) Streptococus nigricans
    c) Pneumococous pneumoniae d) Bacillus subtilis
83. Staphylococcus has
    a) Cubical colony b) Bunch – like irregular colony
    c) Chain like colony d) Plate like colony
84. Antibiotic flavicin (Neomycin) is obtained from
    a) Aspergillus fumigatus b) Aspergillus clavatus
    c) Streptomyces griseus d) Streptomyces fradiae
85. Which one of the following sets includes the bacterial discuses?
    a) Cholera, typhoid, mumps
    b) Tetanus, tuberculosis, measles
    c) Malarias, mumps, poliomyelitis
    d) Diphtheria, leprosy, plague
86. Streptomycin rimosus in the source of antibiotic?
    a) Chloromycetin b) Erythromycin
    c) Aureomycin d) Terramycin
87. The folds of plasma membrane in bacterial cells are known as
    a) Episomes b) Mesosomes
    c) Spherosomes d) Acrosomes
88. Bacteria were discovered by
    a) Robert Koch b) Robert Hooke
    c) A.V. Leeuwen hock d) Louis Pasteur
89. Some bacteria are not easily killed because of
    a) Their tolerant power b) Chitinous wall
    c) Capsule and endospore formation d) Mesosomes
90. A sexual reproductive structure in bacteria is known as
    a) Akinate b) Heterocyst
    c) Endospore d) Excospore

91. Which one belongs to monera?
    a) Amoebe b) Escherichia
    c) Gelidium d) Spirogyra
92. Currently bacteria are included in
    a) Thallophyta b) Mycota
    c) Monera d) Protista
93. A protein rich organism is
    a) Spirulina /Nostoc b) Chlamydomonas
    c) Ulothrix/spirogyra d) Oedogonium
94. Smallest organisms capable of growth, division and reproduction are
    a) Bacteria b) Viruses
    c) Mycoplasma d) Actinomycetes
95. Vibrio cholera is like as
    a) Spring b) Comma
    c) Sphere d) Rod
96. The stored food in blue-green algae is
    a) Starch b) Glucose
    c) Cellulose d) Related to glycogen
97. Diphtheria is caused by
    a) Diplococcus b) Corynebacterium
    c) Bacillus d) Vibrio
98. An obligate anaerobe is
    a) Ulothrix b) Spirogyra
    c) Methane bacteria d) Onion
99. Gram stain represents
    a) A technique for staining bacteria and developed by Gram
    b) A stain got from gram
    c) A cytochemical technique for differentiation of mitochondria
    d) A trade names
100. Bacterial size is
    a) 2-10 μm b) 10-15 μm
    c) 100-200 μm d) 15-50 μm

101. Name the organism which do not derive energy directly or indirectly from sun
   a) Chemosynthetic bacteria b) Pathogenic bacteria
   c) Symbiotic bacteria d) Mould
102. Bacteria having comma like single curve in their body are
   a) Bacilli b) Cocci
   c) Vibrio d) Spirilla
103. Bacteria generally multiply vegetatively through
   a) Conidia b) Binary fission
   c) Budding d) Multiple fission
104. Sexual reproduction does not occur in
   a) Nostoc b) Riccia
   c) Ulothrix d) Rhizopus
105. In bacteria, sex is determined by presence of
   a) Pili b) Episome
   c) Mesosome d) Flagella
106. Streptococcus lactis is responsible for
   a) Conversion of curd into lactic acid
   b) Tanning of leather
   c) Conversion of milk into curd
   d) Flavoring of tea
107. Streptomyces is a member of
   a) Viruses b) Fungi
   c) Eubacteria d) Actinomycetes
108. Vinegar is produced by
   a) Two-step process first fermentation of sugar by yeast, second fermentation of ethyl alcohol by acetic acid bacteria
   b) Fermentation of sugar by Lactobacillus
   c) Fermentation of sugar by Aspergillus
   d) Fermentation of sugar by Saccharomyces cerevisiae
109. Botulism is
   a) Human disease due to parasitic bacteria b) Disease of various organisms
   c) A type of food poisoning d) A viral disease

110. Vinegar is fermented from alcohol by
   a) Azotobacter b) Clostridium
   c) Acetobacter acetic d) Bacillus subtitlis
111. The main difference in Gram (+ ve) and gram (- ve) bacteria resides in their
   a) Cell wall b) Cell membrane
   c) Cytoplasm d) Flagella
112. Gram (-)ve bacteria differ from Gram (+)ve bacteria in having
   a) Thick wall b) Complex wall
   c) Simple wall d) Absence of wall lipids
113. Pili are employed by bacteria for
   a) Locomotion b) Sexual contact
   c) Asexual reproduction d) Location of pray
114. Which changes proteins into ammonia?
   a) Rhizobium b) Nitrobacter
   c) Azotobacter d) Bacillus mycodis
115. Bacterial cell wall is composed of
   a) Lipid b) Cellulose
   c) Chitin d) Mucopeptide / peptidoglycan
116. Largest population of organisms of any type on earth is of
   a) Insects b) Algal
   c) Bacteria d) Fungi
117. Osmotrophs are
   a) Bacteria b) Fungi
   c) Both (a) and (b) d) Algae
118. Bacteria lacking flagella and moving by gliding are
   a) Rickettsiae b) Eubacteria
   c) Spirochaetes d) Myxobacteria
119. Milk is spoiled / fermented / curdled by
   a) Rhizobium b) Lactobacillus
   c) Azotobacter d) Clostridium

120. Some bacteria produce resting spores during unfavorable conditions. They are
   a) Exospores b) Endospores
   c) Aplanospores d) Chlamydospores
121. E. coli is used extensively in biological research as it is
   a) Easily cultured b) Easily available
   c) Easily to handle d) Easily multiplied in host
122. Sterilization by autoclaving is carried out to
   a) Kill bacteria & other pathogens b) Kill viruses
   c) Kill bacteria and enzymes d) Inactivate enzyme
123. Bacteria differ from viruses in
   a) Pathogenic nature b) Genetic material
   c) Having well defined cytoplasm d) Lacking proper nucleus
124. The smallest living cells with cell wall are
   a) Mycoplasma b) Viroids
   c) Blue green algae d) Bacteria
125. Many bacteria possess hairy appendages on their cell wall. They are
   a) Hairs b) Flagella
   c) Cilia d) Fimbrial
126. In bacterial culture, glassware and nutrients are sterilized through
   a) Water bath at 200^{0}C b) Dry air oven at 200^{0}C
   c) Dehumidifier d) Autoclave
127. Antibiotics are
   a) Pesticides b) Bactericide
   c) Herbicides d) Macro biocides
128. Chain of rod-shaped bacteria is
   a) Streptococcus b) Streptobacillus
   c) Staphylococcus d) Staphylobacillus
129. Food poisoning is due to
   a) Clostridium botulinum b) Salmonella typhi
   c) Escherichia coli d) Bacillus megatherium

130. Nuclear material of bacterial cell is known as

a) Nucleus b) Nucleolus

c) Plasmid d) Nucleoid

131. The condition of having a single flagellum at one end of a bacterium is

a) Peritrichous b) Amphitrichous

c) Lophotrichous d) Monotrichous

132. Food can be unspoiled at

a) High temperature b) Low temperature

c) Osmotic temperature d) All the above

133. Gram (+) and Gram (-) forms of bacteria are differentiable through staining with

a) Saffranin + Gentian violet b) Soffranrin + iodine

c) Acetocarmine + iodine d) Crystal violet + iodine

134. A bacterium divides every 35min. If a culture containing 10^5 cell/ml is grows for 175 min, what will be cell concentration that period?

a) 175×10^5 b) 35×10^5

c) 5×10^5 d) 32×10^5

135. A few organisms are known to multiply at 100^0- 108^0C. They are probably

a) Thermophiles sulphur bacteria b) Thermophiles subaerial fungi

c) Hot spring blue green algae d) Marine archae bacteria

136. Botulism caused by clostridium botulinum affects

a) Spleen b) Intestine

c) Neuromuscular junctions d) Lymph glands

137. Bacterial DNA is

a) Straight b) Helical

c) Membrane bound d) Circular and free

138. Fruits, meat and milk are preserved at room temperature through the process of

a) Pasteurization b) Freezing

c) Dehydration d) Vernalization

139. Streptomycin is effective against bacteria

a) Gram (+) b) Gram (-)

c) Gram neutral d) Both gram (+) and gram (-)

140. Bacteria can prepare food from

a) NO_3 b) N_2

c) O_2 d) Glycogen

141. A bacterium having flagella on the opposite ends is

a) Monotrichous b) Lophotrichous

c) Amphitrichous d) Polytrichous

142. All bacterial cells get stained with

a) Mercuric chloride b) Crystal violet

c) Crystal violet + Iodine d) Sofranin

143. Peritrichous bacteria have flagella

a) All over the body b) At one end

c) All both end d) None

144. The bacterium (clostridium botulinum) that causes botulism is

a) Obligate aerobe b) Facultative aerobe

c) Facultative anaerobe d) Obligate anaerobe

145. Cell wall of Gram (+) bacteria is formed of

a) Cell wall b) Murein

c) Cellulose and lipid d) Lipid and protein

146. Bacteria that survive high salt concentration and temperature are

a) Cyanobacteria b) Archaebacteria

c) Eubacteria d) Actinomycetes

147. Inner wall of Gram (-) bacteria is formed of

a) Lipoprotein b) Mucopeptide

c) Chromoprotein d) Glycoprotein

148. Shape of staphylococcus bacteria is

a) Circular b) Oval

c) Elongated d) Cubical

149. Thermal bacteria survive in

a) Hot water near 100^0C b) Hot water near 85^0C

c) Hot sulphur spring near 70^0C d) All the above

150. Smallest bacterium is

a) Dia lister b) Nitrosomonas

c) Bacillus d) Spirillium

151 Which are is peritrichous?

a) Pseudomonas b) Bacillus typhosus

c) Spirllium d) Vibrio

152 Heating food and water will check diseases except

a) Salmonella infection b) Hepatitis-B

c) Cholera d) Botulism

153 A bacterium becomes resistant to antibiotic except by

a) Making enzyme for drug degradation

b) Developing impermeability to drug

c) Modification of drug

d) Moving away from drug

154 Bacterial endotoxin is

a) Lipopolysaccharide over the surface b) Protein inside the cell

c) An excreted protein d) None of the above

155 Which is absent in bacteria

a) Cell wall b) Mesosome

c) Mitochondria d) RNA

156 Gram negative bacteria detect and respond to chemicals by

a) Muramic acid b) Porins

c) Lipopolysaccharides d) Volutin granules

157 Amino acid found only in bacteria and cyanobacteria is

a) Muramic acid b) Methionine

c) Glutamic acid d) Diaminopimelic acid

158 Which is source of vitamin B_{12}?

a) Pseudomonas b) Spirulina

c) Nostoc d) Oscillatoria

159 Bacterial flagella are formed of

a) Amines b) Proteins

c) Lipids d) Carbohydrates

160 Most of the bacteria reproduce by

a) Budding b) Sexually

c) Binary fission d) Sporulation

161. A chain of coccoid bacterial cells is
   a) Staphylococcus
   b) Monococcus
   c) Diplococcus
   d) Streptococcus

162. Bacteria are considered plants as they
   a) are green in color
   b) have rigid cell wall
   c) have chlorophyll
   d) have stomata

163. Disease associated with secretions of toxin is
   a) Tetanus
   b) T.B.
   c) Food poisoning
   d) AIDS

164. Curing of tea leaves is brought about by the activity of
   a) Fungi
   b) Bacteria
   c) Viruses
   d) Mycorrhiza

165. When a bacterial cell possesses a flagellum on its anterior and posterior, the condition is called
   a) Peritrichous
   b) Lophotrichous
   c) Amphitrichous
   d) Monotrichous

166. Bacteria that are smallest in size are
   a) Bacilli
   b) Cocci
   c) Spirilla
   d) Vibrios

167. Penicillin has inhibitory affect over bacteria by
   a) Destruction of nucleus
   b) Inhibition of cell wall synthesis
   c) Stopping entrance of antibody
   d) None

168. F-Factor occurs in
   a) Plasmid
   b) Cosmic
   c) Golgi body
   d) Cell wall

169. Nitrite is changed to nitrate by
   a) Nitrobacteria
   b) Nitrosomonas
   c) Clostridium
   d) Pseudomonas

170. Rod shaped bacteria are called
   a) Cocci
   b) Bacilli
   c) Spirilli
   d) Vibrios

171. Amphitrichous flagellation has

a) Flagella absent b) Flagella at one end

c) Flagella at both the ends d) Flagella all round

172. In bacteria, cell division is

a) Amitotic b) Mitotic

c) Meiotic d) All the above

173. Which causes absorption?

a) Viruses b) Bacteria

c) Mycoplasma d) Chlamydia

174. Which amino acid is present only in bacteria and cyanobacteria?

a) Glycine b) Tyrosine

c) Glutamic acid d) Diaminopimellic acid

175. Spirulina belongs to the kingdom

a) Monera b) Protista

c) Plantae d) Fungi

176. The science of study of bacteria in known as bacteriology. The name 'bacteria' was first of all given by

a) Pasteur b) Alexander flaming

c) Ehrenberg d) Robert Koch

177. Bacteria are found everywhere, expect

a) Cold water b) Soil

c) Boiled water d) Body of man

178. The cell wall of bacteria is made up of

a) Polysaccharides b) Lipids

c) Proteins d) All of the above

179. Bacteria which are spherical are called

a) Neisseria b) Sarcinae

c) Cocci d) Spirilli

180. Bacteria which are highly spiral in shape are called

a) Springeria b) Spirilli

c) Spirochaetes d) Cork-screw bacteria

181. Usually, bacteria reproduce by
   a) Fission b) Conjugation
   c) Transduction d) All the above
182. Sometimes the bacterial cell is enclosed in a
   a) Cell wall of cellulose b) Capsule or slime larger
   c) Plasmalemma d) Cell membrane
183. When milk or other products are heated for about half an hour at about 62^0C and then cooled, the process is called
   a) Pasteurization b) Sterilization
   c) Hydrolysis d) Plasmolysis
184. Nitrifying bacteria play an important role by converting
   a) Ammonia into nitrates
   b) Nitrates into nitrites
   c) Atmosphere nitrogen into nitrites
   d) Decaying plants and animals into ammonia
185. Doctor usually boil their rings and other surgical instruments before use
   a) To remove dust from them b) To clean them
   c) To sterilize them d) It is customary
186. Pasteurized milk is
   a) Free from pathogenic bacteria
   b) Free form bacteria
   c) Sterile and will take longer time to become sour
   d) None of the above
187. Pneumonia is caused by
   a) Virus b) Bacteria
   c) Fungi d) Algae
188. Typhoid in man is caused by
   a) Bacteriophage b) Eberthella typhosa
   c) Penicillium d) Chlorella
189. Of all organisms, the most adaptable and versatile are
   a) Virus b) Bacteria
   c) Algae d) Fungi

190. Cocci type of bacteria are

a) Ciliated
b) Nonciliated
c) Both type
d) None to these

191. Diphtheria is caused by

a) Virus
b) Bacteria
c) Algae
d) Fungi

192. The bacterial decomposition of nitrogenous organic compounds in the absence of abundant oxygen usually results in the formation of substances of offensive odour, chiefly sulphar compounds, such as anaerobic decomposition is called

a) Denitrification
b) Nitrification
c) Nitrogen fixation
d) Putrefaction

193. The decomposition of organic compounds in the presence of O_2 and without the development of odoriferous substance is called

a) Decay
b) Nitrification
c) Nitrogen fixation
d) Denitrification

194. Retting is a process, by which bacteria bring about the

a) Curdling of milk
b) Synthesis of vitamin B_6
c) Nitrogen fixation
d) Separation of fibres of coconut, husk and flax

195. Which scientist in 1884 developed a valuable method of staining the bacteria?

a) Francois Jacob
b) Louis Pasteur
c) Hans Christian Gram
d) James Graham

196. Substances released by some bacteria to cause toxicity in host are

a) Endotoxins
b) Exotoxins
c) Antibiotics
d) None of these

197. The estimated length of DNA molecule in E coli is

a) 1000μ
b) 100μ
c) $100A^0$
d) $1000A^0$

198. Gram stain is used to

a) Determine acidity or alkalinity
b) Identify viruses
c) Identify type of bacteria
d) Detect nucleolus DNA

199 Substance obtained from microorganism and used to inhibit the growth of other microorganisms are

a) Antibiotics b) Antigens

c) Antiseptics d) Antibodies

200 Which is the smallest bacterium

a) Spirillum volutans b) Dialister pneumosintes

c) Salmonella typhosa d) Vibrio cholera

201 The largest known bacterium is

a) Spirillum volutans b) Bacillus anthracis

c) Clostridium botulinum d) Epulopicium fishelsoni

202 Antibiotics cure the disease by

a) Competitive inhibition

b) Removing pain

c) Fighting against organisms causing disease

d) Turning out disease causing organism

203 Most resistance spore in bacteria is

a) Endospore b) Conidian

c) Gonidium d) Oidium

204 Usually, bacteria do not require light for synthesis of food?

a) They may synthesize their food in the absence of light

b) They do not like sunlight

c) The do not synthesize food because of absence of chlorophyll

d) They utilize other electromagnetic radiation

205 Bacteria growing on synthetic media are called

a) Thallus b) Spore

c) Colony d) Tissue

206 Acetobacter is useful in

a) Nitrogen fixation b) Souring of milk

c) Curing of tobacco d) Vinegar industry

207 Diaminopimelic acid is present in the cell wall of

a) Fungi b) Bacteria

c) Red algal d) Some bryophytes

208. The bacterial cell wall is made by up of a substance

a) Glycocalyx b) Slime layer
c) Peptidoglycan d) Dextran

209. During production of tea, the flavoring of leaves is done with the help of

a) Bacillus subtilis b) Micrococcus candidans
c) B. megathenium d) B. radicicola

210. Structurally, the flagella of bacterium are unique in the sense that they are

a) Monofibrillar b) Show 9+2 arrangement
c) Multi fibrillar d) None of them

211. Bacteria with cilia or flagella all round their body are called

a) Lophotrichous b) Amphitrichous
c) Peritrichous d) Monotrichous

212. Fermentation is an anaerobic production of

a) Protein and acetic acid
b) Alcohol, lactic acid and similar compounds
c) Ethers and acetone
d) Alcohol and lipoprotein

213. Who discovered sexual reproduction in bacteria for first time?

a) Leeuwen hock b) Lederberg and Tatum
c) Waksman d) Watson and Crick

214. Putrefying bacteria act upon

a) Fats b) Carbohydrate
c) Proteins d) Starch

215. Name the process of killing all microorganisms

a) Immunization b) Pasteurization
c) Sanitation d) Sterilization

216. Who cultured bacteria for the first time?

a) Robert Koch b) L. Pasteur
c) A.V. Leeuwenhoek d) Waksman

217. To sterilize glassware and the culture medium, which equipment is used

a) Auto calve b) Incubator
c) Dry oven d) Dehumidifier

218. Bacteria were first discovered by

a) A.V. Leeuwenhock b) Robert Hooke

c) Robert Koch d) Louis Pasteur

219. Which of the following statement is correct?

a) All bacteria are autotrophic

b) All bacteria are heterotrophic

c) Mostly bacteria are heterotrophic but some are autotrophic

d) All bacteria are photosynthetic

220. Bacteria are considered to be plants because, the

a) cannot move b) have a rigid cell wall

c) can multiply by fission d) are present every where

221. The bacteria, which are smallest in size are

a) Cocci b) Vibrio

c) Bacilli d) Spirilla

222. The antibiotic is obtained from

a) Mucor b) Ulothrix

c) Streptomyces d) Gonidium

223. The bacteria that can spoil unprocessed canned food are

a) Psychrophiles b) Mesophiles

c) Thermophiles d) None of these

224. All the following human disease are caused by spirochetes except

a) Syphilis b) Lyme disease

c) Food poisoning d) Tetanus

225. Psychrophilic bacteria are those bacteria that grow best

a) In oxygen free environments b) At pH levels of 10 or above

c) At 0^0C to 20^0 C temperature d) Only in the presence of viruses

226. All the following human diseases are due to members of the genus clostridium except

a) Botulism b) Tetanus

c) Severe abdominal pain and diarrhea d) Tuberculosis

227. Bacteria that are pathogenic in the body grow at temperatures
   a) Over 80^0C
   b) At body temperature
   c) At the same temperature as viruses
   d) At mesophilic temperature

228. Halophiles are the bacteria that requires for growth
   a) Extremely hot springs
   b) Extremely strong brine
   c) Arctic areas
   d) Marshy places

229. The penicillin family of antibiotics is used primarily against
   a) Virus
   b) Fungi
   c) Gram +ve bacteria
   d) Gram –ve bacteria

230. Streptomycin is obtained from
   a) Streptomyces griseus
   b) Streptococcus
   c) Bacillus polymyxa
   d) Streptomyces rimosus

231. The similarity between Cyanobacteria and bacteria
   a) Plasmid
   b) Nucleus
   c) Centro some
   d) DNA

232. Which one of the following temperatures would most likely killed a mesophile?
   a) -50^0C
   b) 0^0C
   c) 37^0C
   d) 60^0C

233. Some bacteria cannot kill easily because of it manufactures its own food material.
   a) Their tolerance
   b) Their chitinous cell wall
   c) Capsule presence
   d) Endospore formation

234. Pasteur is famous for his
   a) Work on yeast and fermenting wine
   b) Work a photosynthesis
   c) Germ theory of diseases
   d) Antibiotic discovery

235. An organism that lacks plastids, even than it manufactures its own food material:
   a) Bacteria
   b) Virus
   c) Fungus
   d) Bacteriophage

236. Some diseases caused by bacteria are

a) Measles, mumps, malaria

b) Tetanus, typhoid, tuberculosis

c) Smallpox, sleeping sickness, syphilis

d) Pneumonia, poliomyelitis, psittacosis

237. Diaminopimelic acid and muramic acid are constituents of cell wall in

a) Higher plants b) Fungi

c) Bacteria d) Bacteriophage

238. Wine turns sour because of

a) Heat b) Aerobic bacteria

c) Anaerobic bacteria d) Exposure to light

239. Milk is converted to curd by

a) Xanthomonas citri b) Bacillus magathesium

c) Acetobacter aceti d) Streptococcus lactic

240. The structure formed by the bacterial genome is generally called as

a) Nucleus b) Nucleoside

c) Nucleolus d) Nucleoid

241. Botulism is a

a) Type of food poisoning caused by saprophytic clostridiums

b) Disease caused by staphylococcus in man

c) Disease of citrus

d) Water borne disease of animals

242. Circular DNA is seen in

a) Higher plants b) Bacteria

c) Fungus d) Viruses

243. Botulinum toxin effects

a) Neurotransmitter b) Protein synthesis

c) Pumping of electrolytes & water d) Lysosomal membrane

244. Which statement is correct about E. coli?

a) Gram negative bacteria b) Facultative anaerobe

c) Mesophilic and peritrichous d) All the above

245. The manufacturing of almost milk products such curd, butter, cheese, ghee is based on

a) Bacterial activity
b) Fungal activity
c) Environmental activity
d) None of them

246. The butter on heating is converted into

a) Curd
b) Cheese
c) Yoghurt
d) Ghee

247. First step in the identification of bacteria in food is

a) Microscopic
b) Macroscopic
c) Naked eye
d) None

248. The presence of capsules in bacteria may account for

a) Sliminess
b) Ropiness
c) Haziness
d) Both (a) and (b)

249. The capsule serves to increase the resistance of bacteria against

a) Heat
b) Cold
c) Chemicals
d) Both (a) and (c)

250. Most capsules of bacteria made from polysaccharides i.e.

a) Dextrin
b) Dextran
c) Levan
d) All the above

251. The bacteria oxidize ethyl alcohol to acetic acid and found on fruits and vegetables

a) Acetobacter
b) Aeromonas
c) Bacillus
d) Altromonas

252. A definite spoilage problem in alcohol beverage is caused by

a) Aeromonas
b) Acetobacter
c) Alcaligens
d) Arthrobacter

253. The bacteria are gram negative rods with an optimum temperature for growth of 22 to 28^{0}C

a) Acetobacter
b) Aeromonas
c) Alcaligenes
d) Arthrobacter

254. The bacteria cause ropiness in milk is

a) Alcaligenes viscolactis
b) Alcaligenes metalcaligenes
c) Bacillus subtilis
d) Clostridium botulinum

255. American Type culture collection (ATCC) is related to
   a) Testing of microorganism b) Tissue culture
   c) Fermentation d) Food legislations
256. Which one can spoil a wide variety of meats and meat products?
   a) Brochotrix b) Acetobacter
   c) Bacillus d) Erwinia
257. Campylobacteria are
   a) Oxidase and catalase positive
   b) Gram negative
   c) Curved, S-shaped or spiral shaped
   d) All the above
258. Clostridium bacteria are
   a) Catalase negative b) Acid forming
   c) Gaseous spoilage of canned vegetable d) All the above
259. Erwinia carotovora sub sp. atroseptica produces a black rot in
   a) Carrot b) Redish
   c) Potato d) Sugar beet
260. Erwinia carotovora sub sp. betavasculorum cause soft rot in
   a) Sugar beets b) Carrot
   c) Potato d) Redish
261. Flavobacterium may cause discolourtion on the surface of
   a) Meats b) Potato
   c) Mushroom d) Biscuits
262. Gluconobacter causes ropiness in
   a) Jam b) Jelly
   c) Curd d) Beer

ANSWER KEY

1	**2**	**3**	**4**	**5**	**6**	**7**	**8**	**9**	**10**
a	c	a	a	c	c	c	d	c	d
11	**12**	**13**	**14**	**15**	**16**	**17**	**18**	**19**	**20**
a	b	b	b	c	a	d	a	a	c

21	22	23	24	25	26	27	28	29	30
c	c	a	a	a	b	a	a	c	a
31	**32**	**33**	**34**	**35**	**36**	**37**	**38**	**39**	**40**
a	c	d	d	d	d	b	d	c	d
41	**42**	**43**	**44**	**45**	**46**	**47**	**48**	**49**	**50**
d	b	b	c	a	d	d	a	d	d
51	**52**	**53**	**54**	**55**	**56**	**57**	**58**	**59**	**60**
c	b	c	c	b	a	a	d	a	c
61	**62**	**63**	**64**	**65**	**66**	**67**	**68**	**69**	**70**
b	d	a	a	b	c	a	d	d	d
71	**72**	**73**	**74**	**75**	**76**	**77**	**78**	**79**	**80**
d	b	d	b	a	b	b	d	b	a
81	**82**	**83**	**84**	**85**	**86**	**87**	**88**	**89**	**90**
c	d	b	d	d	d	b	c	c	c
91	**92**	**93**	**94**	**95**	**96**	**97**	**98**	**99**	**100**
b	c	a	c	b	d	b	c	a	a
101	**102**	**103**	**104**	**105**	**106**	**107**	**108**	**109**	**110**
a	c	b	a	a	c	d	a	c	c
111	**112**	**113**	**114**	**115**	**116**	**117**	**118**	**119**	**120**
a	b	b	d	d	c	c	d	b	b
121	**122**	**123**	**124**	**125**	**126**	**127**	**128**	**129**	**130**
a	a	c	d	d	d	b	b	a	d
131	**132**	**133**	**134**	**135**	**136**	**137**	**138**	**139**	**140**
d	b	d	d	c	c	d	c	d	d
141	**142**	**143**	**144**	**145**	**146**	**147**	**148**	**149**	**150**
c	b	a	d	b	b	b	a	c	a
151	**152**	**153**	**154**	**155**	**156**	**157**	**158**	**159**	**160**
b	b	d	a	c	c	d	b	b	c
161	**162**	**163**	**164**	**165**	**166**	**167**	**168**	**169**	**170**
d	b	c	b	c	b	b	a	a	b
171	**172**	**173**	**174**	**175**	**176**	**177**	**178**	**179**	**180**
c	a	c	d	a	c	c	d	c	c
181	**182**	**183**	**184**	**185**	**186**	**187**	**188**	**189**	**190**
a	b	a	a	c	c	b	b	b	b
191	**192**	**193**	**194**	**195**	**196**	**197**	**198**	**199**	**200**
b	d	a	d	c	a	a	c	a	b
201	**202**	**203**	**204**	**205**	**206**	**207**	**208**	**209**	**210**
d	a	a	a	c	d	b	c	b	c

211	**212**	**213**	**214**	**215**	**216**	**217**	**218**	**219**	**220**
c	b	b	c	d	a	a	a	c	b
221	222	223	224	225	226	227	228	229	230
a	c	c	c	c	d	d	b	d	a
231	**232**	**233**	**234**	**235**	**236**	**237**	**238**	**239**	**240**
d	d	d	c	a	b	c	b	d	d
241	**242**	**243**	**244**	**245**	**246**	**247**	**248**	**249**	**250**
a	b	a	d	a	d	a	d	d	d
251	**252**	**253**	**254**	**255**	**256**	**257**	**258**	**259**	**260**
a	b	b	a	a	a	d	d	c	a
261	**262**								
a	d								

3

Food Microbiology-II

1. The fungi are usually:
   a) Parasites or saprophytes b) Lithophytes
   c) Epiphytes d) Insectivorous
2. The branch of botany under which fungi are studies is known as
   a) Bacteriology b) Parasitology
   c) Mycology d) Bryology
3. Rhizopus is also known as
   a) Grown gall b) Black mould
   c) Green mould d) Blue mould
4. The fungi are mostly:
   a) Autotrophic b) Holotrophic
   c) Chemotrophic d) Heterotrophic
5. The parasitic fungi take their nutrition from their host with the help of:
   a) Soredia b) Paraphyses
   c) Haustoria d) Fruiting bodies
6. The importance of study of parasitic fungi is that they are:
   a) Used as food b) Ornamental
   c) Causes of diseases d) Condiments
7. The fungi that are used as food are
   a) Slime mould b) Mushrooms
   c) Mildews d) Rust and smut
8. Rhizopus belongs to
   a) Phycomycetes b) Ascomycetes
   c) Basidiomycetes d) Deuteromycetes

9. Rhizopus and yeast are
   a) Autophytic b) Epiphytic
   c) Parasitic d) Saprophytic
10. Which is an edible fungus
   a) Agaricus b) Rhizopus
   c) Puccinia d) Smut
11. Rhizopus, which is saprophytic by nature, is known as
   a) Rust b) Smut
   c) Bread mold d) Pond sills
12. The mycelium of Rhizopus is
   a) Aseptate and Binucleate b) Aseptate and multinucleate
   c) Septate and nucleate d) None of above
13. Reproduction of Rhizopus is
   a) Only sexual b) Only asexual
   c) Vegetative, asexual & sexual d) None of the above
14. The sexual reproductive orgasm in Rhizopus is
   a) Sporangiophore b) Columella
   c) Aplanogametangium d) None
15. The Rhizopus in
   a) Homothallic b) Heterophillic
   c) Both (a) and (b) d) None
16. The Rhizopus gametes are formed in
   a) Mycelium b) Sporangiophores
   c) Suspensor cells d) Gametangia
17. Yeast belongs to class
   a) Deuteromycetes b) Ascomycetes
   c) Zygomycetes d) Basidiomycetes
18. The chief characteristics of class Ascomycetes is
   a) Formation of spores b) Formation of zoospores
   c) Hypes d) Formation of ascospores
19. Yeast grows more rapidly in
   a) Sugar solution b) Saline water
   c) Cow dung d) Distilled water

20. Yeast is different from Rhizopus and penicillin is being
    a) Multicellular b) Unicellular
    c) Acellular d) In having unseptate hyphae
21. The most common method of vegetative reproduction in yeast is
    a) By ascogonium b) By budding
    c) By forming ascospores d) By fragmentation
22. The most important property of yeast is
    a) Fermentation b) Distillation
    c) Reduction d) Oxidation
23. Penicillin was discovered by:
    a) Alexander Fleming b) Robert Koch
    c) S. Wasksman d) C.J. Alexopoulos
24. A characteristic spore of Agaricus, a mushroom is known as
    a) Conidium b) Basidiospore
    c) Ascospore d) Pycniospore
25. Aflatoxin is produced by
    a) Clostridium botulinum b) Aspergillus flavus
    c) Vibrio coma d) Agaricus bisporus
26. Metalue are found in
    a) Penicillium b) Aspergillus
    c) Agaricus d) Saccharomyces
27. For the production of which microbial action is not required?
    a) Wine b) Curd
    c) Cheese d) Casein
28. Facultative parasite is
    a) Parasite but may function as saprophyte
    b) Always parasite
    c) Always saprophyte
    d) Saprophyte but may function as a parasite
29. Who discovered streptomycin?
    a) A. Fleming b) S. Wasksman
    c) J.S. Wasson d) Alexopoulos

30. What is streptomycin?
    a) A harmone b) An enzyme
    c) An antibiotic d) A virus
31. Yeast is the source of
    a) Vitamin C b) Riboflavin
    c) Sugars d) Proteins
32. The fungi that grow on wood are called
    a) Epibiotic b) Eucarpic
    c) Epixylic d) Epigeon
33. Penicillin is obtained from
    a) Panicillium chrysogenium b) Penicillium notatum
    c) From both (a) and (b) d) None of then
34. Spores formed at the-tip of fungal hype are called
    a) Sporangia b) Arthospore
    c) Conidia d) Zoospores
35. Zoo gametes are not found in
    a) Rhizopus b) Funaria
    c) Fern d) Cycas
36. Discover of Penicillin was based on
    a) Biological antagonism b) Struggle for existence
    c) Struggle for genotype d) None of the above
37. The wall of Rhizopus hypae is composed of
    a) Cellulose b) Chitin
    c) Pectin d) Hemicelluloses
38. The mode of nutrition of Rhizopus, yeast and Penicillin is
    a) Parasitic b) Saprophytic
    c) Symbiotic d) Autotropic
39. The hypae of Rhizopus are
    a) Unbranched, aseptate and uninucleate
    b) Branched, aseptate and uninucleate
    c) Unbranched, septate and multinucleate
    d) Unbranched, septate and coenocytic

40. Thread like filaments which from the plant body of fungi are
    a) Rhizoids b) Paraphyses
    c) Hyphae d) Haptera
41. Yeast like budding of Oidia in Mucor / Rhizopus is called
    a) Palmella b) Chantransia
    c) Torula d) Gongrosira
42. The cell wall of yeast is composed of
    a) Cellulose b) Pectose
    c) Pectin d) Chitin & Mannan
43. Fungus used for the fermentation of cheese is
    a) Mucor mucedo b) Rhizopus nigricans
    c) Penicillium camemberti d) Penicillium chrysogenum
44. Yeast cell divides by
    a) Mitosis only b) Mitosis and Amitosis
    c) End mitosis and Amitosis d) Mitosis and Endo mitosis
45. Yeast is economically important because they
    a) Spread plant diseases
    b) Spread animal disease
    c) Are used in tea and tobacco industry
    d) Are used in wine and baking industry
46. Yeast differs from Rhizopus in being
    a) Multicellular and coenocytic b) Unicellular and uninucleate
    c) Unicellular and coenocytic d) Filamentous
47. When a moist bread is kept exposed in air, it becomes moldy and black because
    a) Spores are present in the water
    b) Spores are present in the bread
    c) Spores are present in the air
    d) The bread decomposes
48. The average diameter of the yeast cell is
    a) 2 μm b) 4 μm
    c) 10 μm d) 20 μm

49. Yeast is abundantly found in
    a) Moist bread
    b) Horse dung
    c) Organic substance rich in sugar
    d) Organic substance rich in fats

50. The nutrition in yeast is saprotrophic. They have the power to convert sugar solution into alcohol by secreting on enzyme complex known as
    a) Amylase b) Zymase
    c) Invert age d) Kinase

51. Penicillin is commonly known as
    a) White mould b) Blue green algae
    c) Black mould d) Yellow mould

52. The edible fungal are
    a) Rusts b) Moulds
    c) Mildews d) Mushroom

53. Penicillin is economically important because it
    a) cause pathogenic disease b) spoils food material
    c) used in preparation of alcohol d) provides antibiotic drugs

54. Amanita, a poisonous fungus is a
    a) Mushroom b) Bracket fungus
    c) Toad stool d) Puffball

55. The important antibiotic wonder drug extracted from Penicillium is
    a) Penicillin b) Aureomycin
    c) Streptomycin d) Terramycin

56. Penicillin is produced by
    a) P. chrysogenum b) P. divaricatum
    c) P. expansum d) P. claviforme

57. Penicillium Roquefort and P. camembert are responsible for
    a) Pathogenic disease in man b) Pathogenic disease in plants
    c) Imparting flavors to cheese and ripe fruits d) None of the above

58. Yeast is not used in the manufacturing of
    a) Toddy b) Penicillin
    c) Wine d) Baking soda
59. The bread becomes soft and porous when the yeast cells are mixed in the lump of dough of wheat flour, because
    a) Yeast is soft and flour also become soft
    b) Yeast produces acetic acid and alcohol which gives softness to the bread
    c) Evolution of CO_2 makes the bread spongy
    d) Yeast produces benzoic acid
60. Which of the following is not a fungus?
    a) Sargassum b) Mucor
    c) Agaricus d) Morchella
61. The fruit juices bitter in taste if they are kept in open place for some time, because:
    a) Some internal factors
    b) Fermentation of the juice by yeast
    c) Bacteria of the atmosphere read with the juices
    d) All the above three statements are correct
62. Vegetative reproduction in yeast takes place of
    a) Akinetes b) Aplanospores
    c) Ascospores d) Budding
63. To digest the food that lies in external medium, a saprophyte secret
    a) Enzymes b) Hormones
    c) Sugar d) None of these
64. Severe famine of West Bengal of 1942-43 was due to destruction of rice crop of a fungal called
    a) Helminthosporium b) Penicillium
    c) Puccinia d) Rhizopus
65. Commercial source of manufacture of citric acid is
    a) Citrus fruits b) Aspergillus
    c) Bacteria d) Penicillium

66. All the following conditions are necessary for the growth of the moulds, mucor or Penicillin, except
    a) Warmth b) Light
    c) Carbohydrate d) Water
67. Aflatoxin is produced by
    a) Virus b) Bacterium/mycobacterium
    c) Fungus Aspergillus flavus d) Nematode
68. Agaricus Mushroom is a
    a) Saprophyte b) Photosynthesizes of food material
    c) Facultative parasite d) Obligate parasite
69. Fungi are always
    a) Autographs b) Heterotrophy
    c) Saprophytes d) Parasites
70. Generally, in laboratory cultures of Rhizopus, there is no formation of zygospore because
    a) There is deficiency of oxygen
    b) There is deficiency of light
    c) Due to the absence of both (+) and (-) strains of mycelia
    d) Presence of (+) and (-) strain of mycelia
71. Yeast is used in the production of
    a) Ethyl alcohol b) Acetic acid
    c) Cheese d) Curd
72. Bread dough rises because of the action of
    a) Heat b) Kneading
    c) Bacteria d) Yeast
73. Fungi can be stained by
    a) Cotton blue b) Safranine
    c) Glycerin d) Lacto phenol
74. Perfect stage of fungus means
    a) When fungus is perfectly healthy
    b) When it produces asexually
    c) When it reproduces sexually
    d) When it forms prefect resting spores

75. Common bread mould is

a) Aspergillus b) Penicillium

c) Erysiphe d) Rhizopus

76. Basidiospores are characteristics of

a) Bread mould b) Mushrooms

c) Aspergillus d) Yeast

77. Red / Pink bread mould is the common name for

a) Neurospora b) Mucor

c) Aspergillus d) Rhizopus

78. Fungi usually store the reserve food material in the form of

a) Starch b) Glycogen oil

c) Lipid d) Protein

79. Fungi can be distinguished from algae by the fact that

a) Mitochondria are absent

b) Cell wall is cellulosic and chlorophyll is absent

c) Nucleus is present

d) Cell wall is chitinous and chlorophyll is absent

80. One of the following is used in the baking of the bread

a) Zygosaccharomyces octosporus

b) Saccharomyces cerevisiae

c) Saccharomycodes ludwigi

d) Rhizopus stolonifer

81. Common form of food stored in fungal cells is

a) Starch b) Sucrose

c) Glucose d) Glycogen

82. Yeast cells are rich in vitamins

a) A b) B

c) C d) D

83. Fungi differ from algae in being mostly

a) Heterotrophic b) Autotrophic

c) Parasitic d) Epiphytic

84. In yeast, during budding, which occurs

a) Synapses b) Unequal division of cytoplasm

c) Doubling of chromosomes d) Spindle formation

85. First antibiotic isolated was

a) Neomycin b) Terramycin

c) Streptomycin d) Penicillin

86. Puffball is

a) Alga b) Fungus

c) Slime mould d) Composite organism

87. Storage grains come to have aflatoxin due to growth of

a) Virus b) Yeast

c) Bacterium d) Aspergillus flavous

88. Penicillin was discovered in the year of

a) 1901 b) 1919

c) 1929 d) 1909

89. Study of fungi is

a) Palynology b) Mycology

c) Phycology d) Microbiology

90. Gills are found in

a) Agaricus b) Puccinia

c) Aspergillus d) Deuteromycetes

91. Alcoholic fermentation is performed by

a) Cholrella b) Agaricus

c) Yeast d) Puccinia

92. Yeast produces an enzyme complex that is responsible for fermentation. The enzyme complex is

a) Aldolase b) Dehydrogenase

c) Invertase d) Zymase

93. An edible fungus is

a) Aspergillus b) Ustilago

c) Polyporus d) Morchella

94. Which one is not a mode of reproduction in yeast?
   a) Budding b) Plasmogamy
   c) Oogamy d) Ascospore formation

95. An organism which lacks sterile covering over its egg is
   a) Yeast b) Funaria
   c) Riccia d) cycas

96. Antibiotic is got from
   a) Mucor b) Gelidium
   c) Ulothrix d) Penicillium

97. In Albuge, the food reserve is mostly
   a) Glycogen b) Volutin granules
   c) Protein granules d) Fat

98. Which are produces alcohol?
   a) Clostridium botulinum b) Leuconostoc citrovorum
   c) Saccharomyces cerevisiae d) Toruplosis utilis

99. Yeast is employed for production of
   a) Curd b) Cheese
   c) Acetic acid d) Ethyl alcohol

100. A chemical substance produced by a micro-organism for inhibiting the growth of another is
   a) Antibody b) Antibiotic
   c) Aflatoxin d) Anti-allergic

101. Common bread mold is
   a) Yeast b) Rhizopus
   c) Clostridium d) Myxo virus

102. Chitin present in fungal wall has a formula of
   a) $(C_{22}H_{54}N_4O_{21})_n$ b) $(C_{21}H_{54}N_4O_{22})_n$
   c) $(C_{22}H_{54}N_4O_{13})_n$ d) $(C_{22}H_{24}N_4O_{21})_n$

103. The fungus without a mycelium is
   a) Rhizopus b) Saccharomyces
   c) Puccina d) Phytophthera

104. Yeast is

a) Purely aerobic b) Anaerobic

c) Rarely anaerobic d) Both aerobic & anaerobic

105. Which one is a laboratory weed?

a) Penicillium b) Aspergillus

c) Neurospora d) Yeast

106. 'Witches Broom' of legumes is due to

a) Mycoplasma b) Bacterium

c) Fungus d) Virus

107. Which is correct about cell wall of bacteria and fungi? Both have

a) Glycopeptide b) N-acetyl glucosamine

c) N-acetyl glucosamine and cellulose d) Chitin

108. Saccharomyces cerevisiae is

a) Akaryote b) Prokaryote

c) Mesocaryote d) Eukaryote

109. Antibiotic flavicin is produced by

a) Aspergillus fumigatus b) A. clavatus

c) Streptomyces griseus d) S. fradiae

110. In yeast, cell wall contains

a) Amylase and glucose b) Glucose and mannose

c) Glucose and muramic acid d) Sucrose and mannose

111. Citric acid is got from

a) Aspergillus niger b) Polyporus species

c) Penicillium notatum d) Sacchromyces cerevisiae

112. Cheese maturation is connected with

a) Aspergillus oryzae b) Aspergillus niger

c) Penicillium camemberi d) Penicillium chrysogenum

113. Aflatoxicosis of poultry is due to

a) Candida albicans b) Penicillium notatum

c) Aspergillus flavus d) Aspergillus fumigates

114. Black colour of bread mould is due to
   a) Zygophores b) Colour of hyphae
   c) Decay of organic matter d) Sporangia
115. Torula' condition' occurs in
   a) Rhizopus b) Ulothrix
   c) Spirogyra d) Riccia
116. Toxin is secreted during storage condition by
   a) Fusarium b) Colletotrichum
   c) Penicillium d) Aspergillus
117. Purified antibiotic penicillium of Penicillium notatum was obtained by
   a) Alexander Fleming b) Howard Florey
   c) Robert Hooke d) Carolus Linnaeus
118. Which statement is correct about moulds?
   a) Involved in spoilage of food
   b) Manufacturing of certain foods
   c) Both above
   d) None
119. Molds are useful in
   a) Ripening of some kind of cheese
   b) Making of oriented foods
   c) Bread making & citric acid production
   d) All of the above
120. Which is the general characteristic of molds?
   a) Multicellular b) Filamentous
   c) Fuzzy or colony appearance d) All the above
121. Rhizoids or hold fasts, an identification characteristic of when molds.
   a) Rhizopus and Absidia b) Aspergillus
   c) Geotrichum d) Mucor
122. Foot cell is found in
   a) Aspergillus b) Mucor
   c) Geotrichum d) All the above

123. Dichotomous or y-shaped branching, an identification character of

a) Aspergillus b) Cladosporium

c) Penicillium d) Geotrichum

124. Molds required moisture for growth

a) Less than bacteria and yeast b) More than bacteria and yeast

c) Unknown d) None

125. The optimal temperature for most growth is around

a) 25-30^{0}C b) 35-37^{0}C

c) -5-10^{0}C d) 10-15^{0}C

126. The molds growth below the refrigeration and freezing is called

a) Mesophilic b) Thermophilic

c) Psychrotrophic d) None

127. What is temperature range of psychotropic molds for growth?

a) -5 to -10^{0}C b) -30 to -37^{0}C

c) 10 to 15^{0}C d) 35 to 37^{0}C

128. Certain chemical compounds are mycostatic, inhibiting the growth of molds, i.e.

a) Sorbic acid b) Propionate

c) Acetates d) All the above

129. Molds do not have

a) Roots b) Stems

c) Leaf d) None

130. Mucor rouxii is used in the "Amylo" process for the saccharification of

a) Fat b) Starch

c) Glucose d) Glycogen

131. Which is called the bread mold

a) Rhizopus stolonifer b) M. rouxii

c) Aspergillus niger d) None

132. Which molds is found on meat in chilling storage, causing whiskers on the meat?

a) Aspergillus niger b) Rhizopus stolonifer

c) Thamnidium elegans d) Absidia

133. Which one of the following microorganisms is used in the preparation of bread?

a) Candida utilis | b) Saccharomyces cevarum
c) Aspergillus niger | d) Saccharomyces cerevisiae

134. Which on the following conditions for the heat resistance of microorganism is correct?

a) Psychrophiles<Mesophiles<Thermophiles
b) Psychrophiles>Mesophiles>Thermophiles
c) Thermophiles >Psychrophiles>Mesophiles
d) Mesophiles<Thermophiles<Psychrophiles

135. Penicillium expansum, the blue green-spored mold causes soft rot of

a) Fruits | b) Vegetables
c) Sugarcane | d) Grain

136. P. camemberti with grayish conidia is useful in the ripening of

a) Camembert cheese | b) Roquefort cheese
c) Chedder cheese | d) Swiss cheese

137. Trichothecium is a

a) Pink mold | b) Gray mold
c) Orange mold | d) Cream colors mold

138. Match the Penicillium molds and their principal characteristics given in the two Colum below

P. P. expansum | 1. Blue contact mold
Q. P. digitatum | 2. Yellowish-green conidia
R. P. italicum | 3. Grayish conidia
S. P. camemberti | 4. Bluish-green conidia
T. P. roqueforti | 5. Blue green-spored mold

a) P-5,Q-1,R-2,S-3,T-4 | b) P-5,Q-2,R-1,S-3,T-4
c) P-1,Q-2,R-3,S-4,T-5 | d) P-5,Q-4,R-3,S-2,T-1

139. Which one of the following is often called dairy mold?

a) Geotrichum candidum | b) Rhizopus
c) Aspergillus niger | d) P. notatum

140. Geotrichum genus mold is not growth in the color of

a) White b) Yellowish

c) Orange or red d) Purple

141. Neurospora (Monilia) sitophila, the most important species in foods, sometimes is termed as

a) Red bread mold b) Dairy mold

c) Bread mold d) None

142. Sporotrichum carnis is found to grow on chilled meats, where is cause

a) White spot b) Black spot

c) Soft rot d) Ripening in cheese

143. Cladosporium herbarum is a dark mold and causes on food

a) Black spot b) White spot

c) Granular surface d) Red spot

144. Which molds is not spoiled the foods?

a) Alternaria b) Fusarium

c) Cladosporium d) Torulopsis

145. The microorganism which are generally not filamentous but unicellular and ovoid or spheroid and which reproduce by budding or fission is called

a) Bacteria b) Virus

c) Yeast d) Molds

146. Yeast fermentation are not involved in the manufacturing of

a) Bread b) Bear

c) Vinegar d) Sauerkraut

147. Yeasts are grown for

a) Foods b) Enzymes

c) Fermentation d) All the above

148. Yeasts are

a) Oxidative b) Fermentation

c) Both above d) None

149. The oxidative yeast may green on the surface of liquid as

a) Film b) Pellicle

c) Scum d) All the above

150. The oxidative yeast is also known as

a) Film yeast b) Surface yeast

c) Fermentative yeast d) None

151. Fermentative yeast usually grown throughout the liquid and produce

a) Carbon dioxide b) Ethyl alcohol

c) Water d) Acetic acid

152. The optimal temperature for growth of most yeast is

a) 25-30^{0}C b) 20-25^{0}C

c) 35-47^{0}C d) 0-40^{0}C

153. The growth of most yeasts is favored by an acid reaction in the vicinity of pH

a) 2-3 b) 4-4.5

c) 4.5-7.0 d) Above 7

154. The best source of energy for yeasts

a) Sugar b) Brine

c) Alcohol d) Starch

155. Carbon dioxide produced by bread yeasts accomplishes the

a) Production of wine b) Leavening of bread

c) Ripening of cheese d) None

156. Yeasts grow in soy sauce with its high content of salt about

a) 3% b) 6%

c) 12% d) 18%

157. Saccharomyces rouxii can grow as film on

a) Brine b) Sugar

c) Jam d) Pickles

158. Debaryomyces yeast is very salt tolerant and can grow on cheese brines with as much as salt about

a) 3% b) 10%

c) 18% d) 24%

159. All bakers', brewers, wine and champagne yeasts are

a) S. Cerevisiae b) Candida utilis

c) Torulopsis sphaerica d) S. mellis

160. Improper handling and storage of cereal grains and oilseeds result in the production of toxins known as aflatoxin which are not generally destroyed by normal cooking process. Aflatoxins are produced by

a) Bacteria b) Protozoa

c) Moulds d) Viruses

ANSWER KEY

1	**2**	**3**	**4**	**5**	**6**	**7**	**8**	**9**	**10**
a	c	b	d	c	c	b	a	d	a
11	**12**	**13**	**14**	**15**	**16**	**17**	**18**	**19**	**20**
c	b	c	c	c	d	b	d	a	b
21	**22**	**23**	**24**	**25**	**26**	**27**	**28**	**29**	**30**
b	a	a	b	b	a	d	d	b	c
31	**32**	**33**	**34**	**35**	**36**	**37**	**38**	**39**	**40**
b	c	c	c	a	a	b	b	c	c
41	**42**	**43**	**44**	**45**	**46**	**47**	**48**	**49**	**50**
c	d	c	a	d	b	c	b	c	b
51	**52**	**53**	**54**	**55**	**56**	**57**	**58**	**59**	**60**
b	d	d	c	a	a	c	b	c	a
61	**62**	**63**	**64**	**65**	**66**	**67**	**68**	**69**	**70**
b	d	a	a	b	b	c	a	b	c
71	**72**	**73**	**74**	**75**	**76**	**77**	**78**	**79**	**80**
a	d	a	c	d	b	a	b	d	b
81	**82**	**83**	**84**	**85**	**86**	**87**	**88**	**89**	**90**
d	b	a	b	d	b	d	c	b	a
91	**92**	**93**	**94**	**95**	**96**	**97**	**98**	**99**	**100**
c	d	d	c	a	d	a	c	d	b
101	**102**	**103**	**104**	**105**	**106**	**107**	**108**	**109**	**110**
b	a	b	d	b	c	b	d	a	b
111	**112**	**113**	**114**	**115**	**116**	**117**	**118**	**119**	**120**
a	c	b	d	a	d	b	c	d	d
121	**122**	**123**	**124**	**125**	**126**	**127**	**128**	**129**	**130**
a	a	d	a	a	c	a	d	d	b
131	**132**	**133**	**134**	**135**	**136**	**137**	**138**	**139**	**140**
a	c	d	a	a	a	a	b	a	d
141	**142**	**143**	**144**	**145**	**146**	**147**	**148**	**149**	**150**
a	a	a	d	c	d	d	c	d	a
151	**152**	**153**	**154**	**155**	**156**	**157**	**158**	**159**	**160**
a	a	b	a	b	d	a	d	a	c

4

Food Microbiology-III

1. According to Pasteur statements which one of the following is true
   a) Living organisms discriminate between stereo isomers
   b) Fermentation
   c) Living organisms doesn't discriminate
   d) Both a and b
2. The light emitted by luminescent bacteria is mediated by the enzyme
   a) Coenzyme 'Q' b) Luciferse
   c) Lactose dehydrogenase d) Carboxylase reductase
3. Salt and sugar preserve foods because they
   a) Make them acid
   b) Produce a hypotonic environment
   c) Deplete nutrients
   d) Produce a hypotonic environment
4. In a fluorescent microscope the objective lens is made of
   a) Glass b) Quartz
   c) Polythene d) None of these
5. Direct microscopic count can be done with the aid of
   a) Neuberg Chamber b) Anaerobic Chamber
   c) Mine ral oil d) Olive oil
6. The image obtained in a compound microscope is
   a) Real b) Virtual
   c) Real inverted d) Virtual inverted
7. Enzymes responsible for alcoholic fermentation
   a) Ketolase b) Zymase
   c) Peroxidase d) Oxidase

8. Father of microbiology is
   a) Louis Pasteur   b) Lister
   c) A.V. Leeuwenhock   d) Robert Koch
9. Compound microscope was discovered by
   a) Antony von   b) Pasteur
   c) Jahnsen & Hans   d) None of these
10. Disease that affects many people at different countries is termed as
   a) Sporadic   b) Pandemic
   c) Epidemic   d) Endemic
11. In electron microscope, what material is used as an objective lens?
   a) Magnetic coils   b) Superfine glass
   c) Aluminum foils   d) Electrons
12. Tuberculosis is a
   a) Water borne disease   b) Air borne disease
   c) Food borne disease   d) Arthropod borne disease
13. In Electron Microscope source of electrons is from
   a) Mercury lamp   b) Tungsten Metal
   c) Both a and b   d) None of these
14. The resolution power of the compound microscope is
   a) 0.2 micron   b) 0.2 millimeter
   c) 0.2 Angstrom Units   d) 0.2 centimeter
15. The capacity of a given strain of microbial species to produce disease is known as
   a) Pathogen   b) Virulence
   c) Infection   d) None of these
16. Electron microscope gives magnification upto
   a) 100 X   b) 2000 X
   c) 50,000 X   d) 2,00,000 X
17. Term Vaccine was coined by
   a) Robert Koch   b) Pasteur
   c) Needham   d) None of these

18. The inventor of Microscope is
    a) Galileo   b) Antony von
    c) Pasteur   d) Koch
19. First Pasteur conducted fermentation experiments in
    a) Milk   b) Food material
    c) Fruit juices   d) Both a and c
20. Compound Microscope was discovered by
    a) A.V. Lewenhoek   b) Pasteur
    c) Janssen and Hans   d) None of these
21. Electron Microscope was discovered by
    a) Prof. Frits   b) Janssen and Hans
    c) Knoll and Ruska   d) None of these
22. Vibrio Cholerae was discovered by
    a) Koch   b) Metchnikoff
    c) John Snow   d) Virchow
23. E. coil was first isolated by
    a) Louis Pasteur   b) Escherich
    c) Shiga   d) Robert Koch
24. Mycobacterium tuberculosis was first discovered by
    a) Robert Koch   b) Edward Jenner
    c) Louis Pasteur   d) None of these
25. Streptococcus pneumonia was isolated by
    a) Robert Koch   b) Edward Jenner
    c) Antony von leewenhock   d) Luis Pasteur
26. Rh factor of the blood was discovered by Scientist
    a) Louis Pasteur   b) Landsteiner and Weiner
    c) Janskey   d) Moss
    e) None of these
27. Rancidity in spoiled foods is due to
    a) Lipolytic organisms   b) Proteolytic organisms
    c) Toxigenic microbes   d) Saccharolytic microbes

28. The bacterial cells are at their metabolic peak during
    a) Lag phase b) Log
    c) Stationary d) Decline
29. Protein particles which can infect are called
    a) Virons b) Prions
    c) Nucleoida d) None of these
30. Endotoxin produced by gram negative bacteria is present in
    a) Peptidoglycan b) Lippolysacharide
    c) Theichoic acid d) Inner membrane
31. A population of cells derived from a single cell are called
    a) Monclonal cells b) Clones
    c) Protoplasts d) Sub culture
32. Hetrolactic acid bacteria produce
    a) Lactic acid only b) Lactic acid + H_2O + CO_2
    c) Lactic acid + CO_2 d) Lactic acid + alcohol + CO_2
33. The viruses that live as parasites on bacteria are
    a) Fungi b) Commensels
    c) Bacteriophages d) None of these
34. Staining material of gram positive bacterium is
    a) Fast green b) Haematoxylon
    c) Crystal violet d) Safranin
35. The pigment present in red algae is
    a) Rhodochrome b) Fucoxanthin
    c) Chlorophyll only d) Chlorophyll + Phycobilin
36. Citrus cancer is caused by
    a) Phytomonas b) Slamonella
    c) Lactobacillus d) Hay bacillus
37. Bacteria that are responsible for fermentation of dairy milk are
    a) Azetobacter b) Rhizobium
    c) Lactobacillus d) Hay bacillus

38. Virus will contain

a) Cell membrane b) Cell wall

c) DNA d) DNA or RNA

39. The wonder drug of second world war is produced by

a) Algae b) Fungi

c) Bacteria d) Plants

40. The enzyme needed in biological systems for joining two molecules is called

a) Lyases b) Diastases

c) Polymerases d) Hydrolase

41. The most infectious food borne disease is

a) Tetanus b) Dysentery

c) Gas gangrene d) Botulism

42. An example for common air borne epidemic disease

a) Influenza b) Typhoid

c) Encephalitis d) Malaria

43. Rancidity of stored foods is due to the activity of

a) Toxigenic microbes b) Proteolytic microbes

c) Saccharolytic microbes d) Lipoytic microbes

44. Virion means

a) Infectious virus particles b) Non-infectious particles

c) Incomplete particles d) Defective virus particles

45. The test used for detection of typhoid fever

a) WIDAL test b) ELISA

c) Rosewaller test d) Westernblotting

46. Bacillus is an example of

a) Gram positive cacteria b) Gram negative bacteria

c) Virus d) Viroid

47. Amoebic dysentery in humans is caused by

a) Palsmodium b) Paramecium

c) Yeast d) Entamoeba histolytica

48. Mordant used in grams staining is
    a) Crystal violet b) Lodine
    c) Saffranin d) All of these
49. Gram staining is an example for
    a) Simple staining b) Differential staining
    c) Negative staining d) None of these
50. Enterotoxin responsible for food poisoning is secreted by
    a) Enterococci b) Entamoeba histolytica
    c) Enterobacteriaceae d) Straphylococci
51. Autolysis is done by
    a) Mitochondria b) Lysosomes
    c) Golgi bodies d) Peroxisomes
52. A facultative anaerobic is
    a) Only grow anaerobically
    b) Only grow in the presence of O_2
    c) Ordinarily an anaerobe but can grow with O_2
    d) Ordinarily an aerobe but can grow in absence of O_2
53. The percentage of O_2 required by moderate anaerobe is
    a) 0% b) < 0.5%
    c) 2 - 8% d) 5 - 10%
54 Pigment bearing structure of bacteria are
    a) Mesosomes b) Plasmids
    c) Mitochondria d) Chromophores
55. Cell wall of gram negative bacteria is
    a) Thick b) Lipids are present
    c) Teichoic acids are absent d) None of these
56. The stain used to demonstrate fungus
    a) Albet b) Nigerosin
    c) Lactophenol cotton blue d) None of these
57. Exotoxina are
    a) Heat labile b) Heat stable
    c) Part of cell wall d) Polymerized complexes

58. The viruses that attack bacteria are
    a) Bacterial viruses b) Bacterial pathogens
    c) Bacteriophages d) Various
59. The size of virus particle may range
    a) 0.02-0.2 μm b) 0.5-10 μm
    c) 0.015-0.2 μm d) 0.1-100 μm
60. Rod shaped bacteria are known as
    a) Cocci b) Comma forms
    c) Bacilli d) Plemorphic froms
61. Thickness of cell wall ranges from
    a) 9-10 nm b) 12-13 nm
    c) 10-25 nm d) 30-40 nm
62. The characteristic shape of the bacteria is maintained because of
    a) Capsule b) Cell wall
    c) Cell membrane d) Slime layer
63. Bacterial capsule is chemically composed of
    a) Polypeptide b) Polynucleotides
    c) Polysaccharides d) Polypeptides or polysaccharides
64. The differences between Gram positive and Gram negative bacteria is shown to reside in the
    a) Cell wall b) Nucleus
    c) Cell membrane d) Mesosomes
65. Bacteria multiply by
    a) Spore formation b) Simple binary fission
    c) Conjugation d) Gametes
66. Bacterial spores are
    a) Weakly acid fast b) Strongly acid fast
    c) Alcohol fast d) Non acid fast
67. The percentage of alcohol used in Gram staining is
    a) 75% b) 90%
    c) 60% d) 25%

68. Gram positive bacteria appear as
    a) Pink b) Violet
    c) Both a & b d) None of these
69. Gram negative bacteria appear as
    a) Pink b) Violet
    c) Both a & b d) None of these
70. The action of alcohol during Gram staining is
    a) Allows the color b) It adds color
    c) Decolorises the cells d) None of these
71. Lipid contents is more in
    a) Gram negative bacteria b) Gram positive bacteria
    c) Same in bot d) None of these
72. Cell-wall is
    a) Thick in Gram positive than Gram negative
    b) Thick in Gram negative than Gram positive
    c) Equal in both
    d) In Gram Negative cell-wall is absent
73. The Lipid content present in Gram positive bacterial cell-wall is
    a) 1-10% b) 1-5%
    c) 2-8% d) None of these
74. Algae means
    a) Fresh water organisms b) Sea weeds
    c) Fresh water weeds d) None of these
75. The study of algae is known as
    a) Algalogy b) Phycology
    c) Mycology d) Bacteriology
76. Alginic acids and its salts are obtained from the wall of
    a) Red algae b) Brown algae
    c) Green algae d) Red and brown algae

77. The molds obtained nutrition from dead and decaying matter which are called

a) Saphrophytes b) Parasites
c) Commensals d) None of these

78. Most molds are capable of growing in the temperature range between

a) 0° - 25°C b) 0° - 35°C
c) 10° - 25°C d) 10° - 35°C

79. The largest virus is

a) Parvo virus b) Pox virus
c) Rhabdo virus d) None of these

80. The smallest virus is

a) Parvo virus b) Rhabdo virus
c) Pox virus d) Adeno virus

81. Shape of bacteriophage is

a) Brick shape b) Bullet shape
c) Helical shape d) Tadpole shape

82. If more than one stain is used, such staining is called

a) Simple staining b) Negative staining
c) Differential staining d) None of these

83. Majority of bacteria are

a) Saprophytes b) Symbionts
c) Commensals d) Parasites

84. Symbionts are

a) Bacteria in symbiotic association
b) The group of fungi in symbiotic association
c) The groups participating ins symbiotic association
d) All of these

85. The best example for symbiotic association is

a) E.coli in intestine of man b) Lichens
c) Normal flora of skin d) All of the above

86. The enzymes responsible for decomposition is
    a) Lipolytic b) Proteolytic
    c) Lysozyme d) Both a & b
87. Attenuation means
    a) Killing of the bacteria (microorganism)
    b) Inactivation of bacteria
    c) More activation th bacteria
    d) Both a & b
88. Main cause for Cholera is
    a) Poverty and insanitation b) Mosquitoes
    c) Toxin produced by pesticides d) None of these
89. AIDS virus is
    a) RNA Virus b) DNA Virus
    c) Retro Virus d) Entero Virus
90. AIDS is cused by
    a) HTLV – I b) Bunya Virus
    c) HTLV – III d) All
91. Screening test for AIDS is
    a) Western blot test b) ELISA test
    c) Both a & b d) VDRL test
92. Confirmatory test for AIDS is
    a) Western blot test b) ELISA test
    c) Karpas test d) Fujerbio test
93. Zihl – Neelson stain is a ………………..
    a) Simple stain b) Counter stain
    c) Differential stain d) None of them
94. Fungi differs with bacteria in that it –
    a) Contain no peptidoglycan b) Are prokaryotic
    c) Susceptible to griseofulvin d) Have nuclear membranes
    e) All of these

95. The dengue fever virus is –
   a) Arbo virus b) Echo virus
   c) Entero virus d) Orthomyxo virus

96. Dengue fever is caused by -
   a) Bacteria b) Virus
   c) Fungi d) Rickettsia

97. The medium used in membrane filter technique was
   a) EMB agar b) EMR-Vp medium
   c) Lactose broth d) Endo agar

98. Lysol is a
   a) Sterilent b) Disinfectant
   c) Antiseptic d) Antifungal agent

99. The method in which the cells are frozen dehydrated is called
   a) Pasteurization b) Dessication
   c) Disinfection d) Lypophilization

100. The technique used to avoid all microorganisms is accomplished by
   a) Sterilization b) Disinfection
   c) Surgical sterilization d) Disinfection sterilization

101. Thermal death time is
   a) Time required to kill all cells at a given temperature
   b) Temperature that kills all cells in a given time
   c) Time and temperature needed to kill all cells
   d) All of the above

102. Temperature required for pasteurization is
   a) Above 150°C b) Below 100°C
   c) 110°C d) None of these

103. Which of the following is ionizing radiation
   a) U.V.rays b) IR
   c) γ rays d) None of these

104. When food material are preserved at a temperature just above freezing temperature, the process is called

a) Freezing
b) Pasteurization
c) Chilling
d) Frosting

105. Which is the following method of sterilization has no effect on spores?

a) Drying
b) Hot air oven
c) Autoclave
d) None of these

106. Autoclaving is carried at

a) Dry heat
b) Atmospheric pressure
c) 120ºC
d) All of these

107. Temperature in pasteurization is

a) 62.8ºC
b) 37.7ºC
c) 68.2ºC
d) 60.8ºC

108. The bacterial culture prepared by pure culture method is

a) Inoculum
b) Suspension
c) Dilution
d) None of these

109. Algae are rich in

a) Carbohydrates
b) Proteins
c) Vitamins
d) All of these

110. Slow freezing requires the conditions

a) 0ºC to 15ºC for 15 min.
b) -6ºC to -10ºC for 10 min.
c) -15ºC to 3 to 72 hrs.
d) None of these

111. Discontinuous heating is called

a) Pasteurization
b) Sterilization
c) Fermentation
d) Tindalisation

112. Isolation is

a) Purification of culture
b) Introduction of inoculums
c) Separation of a single colony
d) To grow microorganisms on surfaces

113. The condition required for autoclave
   a) 121°C temp. and 15 lbs. pressure for 20 min.
   b) 120°C temp. and 20 lbs. pressure for 30 min.
   c) 150°C temp. for 1 hrs.
   d) 130°C temp. for 2 hrs.

114. Lysozyme is effective against
   a) Gram negative bacteria b) Gram positive bacteria
   c) Protozoa d) Helminthes

115. Infrared radiation is method of sterilization by
   a) Dry heat b) Moist heat
   c) Chemical method d) Mechanical method

116. Lyophilization means
   a) Sterilization b) Freeze-drying
   c) Burning to ashes d) Exposure to formation

117. Temperature used for hot air oven is
   a) 100°C for 1 hour b) 120°C for 1 hour
   c) 160°C for 1 hour d) 60°C for 1 hour

118. Phenol co-efficient indicates
   a) Efficiency of a disinfectant b) Dilution of a disinfectant
   c) Purity of a disinfectant

119. This is an agar plate method and is commonly used for estimation of the number of bacteria in milk
   a) Standard plate count (SPC) b) Spread plate
   c) Lawn culture d) Roll tube method

120. Agar is obtained form
   a) Brown algae b) Red algae
   c) Green algae d) Blue- green algae

121. A gram positive organism which produces swarming on culture medium is
   a) Salmonella b) Clostridium
   c) Staphylococci d) Proteus

122. For effective sterilization in an autoclave the temperature obtained is
   a) 50ºC b) 100ºC
   c) 120ºC d) 180ºC
123. Spores are killed by
   a) 70% alcohol b) Clutaraldehyde
   c) Autoclaving d) Both b & c
124. Glassware are sterilized by
   a) Autoclaving b) Hot air over
   c) Incineration d) None of these
125. Tyndallisation was proposed by
   a) Tyndall b) Pasteur
   c) Koch d) Jenner
126. Viruses can be cultivated in
   a) Lab media b) Broth
   c) Living cells d) None of these
127. By pasteurization
   a) All the microorganisms can be removed
   b) Only pathogenic forms can be removed
   c) Only non-pathogenic forms can be removed
   d) All of these are correct
128. The temperature required for pasteurization is
   a) Above 100ºC b) Below100ºC
   c) 100ºC d) None of these
129. In the medium other than nutrients, if any substance is used in excess, that medium is
   a) Enriched medium b) Special medium
   c) Enrichment medium d) None of these
130. Example for indicator medium is
   a) Nutrient Agar b) Nutrient broth
   c) Wilson and Blair d) Czapeck-dox medium

131. Example of Anaerobic medium is
   a) Robertson cooked-meat medium
   b) Nutrient agar
   c) Nutrient broth
   d) Mac-Conkey's agar

132. Best method for getting pure culture is
   a) Streak-plate b) Agar slant
   c) Both a & b d) None of these

133. To transfer cultures from one place to another, the device used is
   a) Slant b) Needle
   c) Inoculation loop d) Autoclave

134. The bacterial culture prepared by pure culture is
   a) Inoculum b) Suspension
   c) Dilution d) None of these

135. Separation of a single colony is
   a) Pure-culturing b) Isolation
   c) Separation d) Both a and b

136. Growth period of the culture is
   a) Inoculation b) Incubation
   c) Incineration d) Isolation

137. At the temperature 160°C for one hour, complete sterilization occur in
   a) Autoclave b) Hot air oven
   c) Laminar flow d) Incubator

138. In autoclave, the principle involved is
   a) Dry heat b) Moist heat
   c) Steam under pressure d) Both b and c

139. The spores of the bacteria which can withstand the moist heat effect also
   a) Bacillus subtilis b) Coxiella burnetti
   c) Bacillus stearothermophilus d) Pseudomonas

140. The sterilization agent is
   a) Ethelene oxide b) Oxygen
   c) Nitrogen d) Carbon tetrachloride

141. Cultures are prepared by penetrating the inoculation loop with suspension into the medium, they are

a) Stock cultures b) Stab cultures

c) Sub-cultures d) None of these

142. The principle involved in the streak plat method is

a) Separation b) Streaking

c) Isolation d) Dilution

143. Culture media for fungi are

a) Potato dextrose agar (PDA) b) Sabouraud's agar

c) Czapekdox agar d) All of the above

144. The term that is used for the bacteria which can withstand pasteurization but does not grow at higher temperatures

a) Thermopiles b) Extreme thermopiles

c) Thermoduric d) Facultative thermopiles

145. A common laboratory method of cultivating anaerobic micro-organisms is

a) Gas pack system b) Brewer jar system

c) Pyrogallic acid over the cotton d) None of these

146. The micro-organisms grow at high salinity are

a) Osmophiles b) Halophiles

c) Bothe a & b d) None of these

147. Non-lactose fermenting colonies seen on Mac Conkey's Medium are

a) Salemonella typhi b) Escherichia coli

c) Klebsiella pneumonia d) Shigella shigae

148. Wilson and Blair medium is used for isolation of

a) Staphylococci b) Salmonella typhosa

c) Vibrio colerae d) Shigella shigae

149. Shigella was first isolated by

a) Shiga b) Schmitz

c) Sonnei d) Robert Koch

150. Which of the following are gas producing Salmonella?

a) S. typhi b) S.entritidis

c) S.cholerasuis d) S.typhimurium

151. Culture medium for Mycobacterium tuberculosis

a) L J medium
b) Mac Conkey's medium
c) Wilson blair medium
d) None of these

152. Slat agar is used for

a) Streptococus
b) Staphylococcus
c) Vibrio
d) Shigella

153. Culture medium used for fungus is

a) Sabouraud's medium
b) Nutrient agar
c) Nutrient broth
d) Minimal agar medium

154. For sterilization of fermentation equipment the method followed is

a) Radiation
b) Chemicals
c) Heating
d) All of these

155. Which virus was first observed?

a) Hepatitis virus
b) TMV
c) Cauliflower mossaic virus
d) None of these

156. Mycotoxins are produced by

a) Bacteria
b) Fungi
c) Algae
d) Protozoans

157. "Agar used for solidifying culture media", which of the following statements is true, except?

a) Melting and solidifying points of agar solution are not same
b) Not add to the nutritive properties of the medium
c) Not affect by the growth of bacteria
d) Semi-solid media is obtained at more than 10% concentration

158. Yeast extract is excellent source of

a) A vitamin
b) Proteins
c) B Vitamin
d) Carbohydrates

159. Biological Oxygen Demand (BOD) is a measure of

a) Industrial wastes poured into water bodies
b) Extent which water is polluted with organic compounds
c) Amount of carbon monoxide inseparably combined with hemoglobin
d) Amount of oxygen needed by green plants during night

160. The following organisms have been proposed as sources of single cell protein
    a) Bacteria b) Yeasts
    c) Algae d) All the three
161. The major constituents in agar are
    a) Fats b) Amino acids
    c) Polysaccharides d) Polypeptides
162. The first phase of a growth curve is
    a) Log phase b) Lag phase
    c) γ phase d) Both a & b
163. Rapid bacterial growth phase is known as
    a) Log b) Lag
    c) Lack d) None of these
164. Mycotoxins are formed during the end of
    a) Lag phase b) Log phase
    c) Death phase d) Stationary phase
165. Bacteria which need oxygen for growth are called
    a) Thermophilic bacteria b) Microaerophilic bacteria
    c) Facultative anaerobic bacteria d) Mycobacteria
166. The most important vitamin for the growth of bacteria is
    a) B-complex b) Vitamin A
    c) Vitamin D d) Vitamin C
167. If the source of energy for bacteria is from chemical compounds they are said to be
    a) Phototrophs b) Autotrophs
    c) Chemotrophs d) Chemolithotroph
168. Vitamin function as
    a) Co-enzymes b) Co-melecules
    c) Building blocks of cell d) None of these
169. The vitamin required for Lactobacillus species is
    a) Riboflavin b) Niacin
    c) Pyridoxine d) Folic acid

170. Vitamin K is necessary for the species

a) Lactobacillus spp. b) Bacillus anthracis

c) Bacteroides melaninogenicus d) All of these

171. The bacteria which are able to grow at 0°C but which grow at 20°C to 30°C, are known as

a) Psychrophiles b) Facultative psychrophiles

c) Average psychrophiles d) Mesophiles

172. In the sigmoid curve (or) growth curve of bacteria how many stages are there

a) 3 b) 4

c) 2 d) 5

173. The reproduction rate is equal to death rate in which stage

a) Decline phase b) Stationary phase

c) Lag phase d) Log phase

174. The organisms which can grow best in the presence of a low concentration of oxygen

a) Aerophilic b) Microaerophilic

c) Aerobic d) Anaerobic

175. Match the following growth characteristics with their respective temperature ranges A to E:

1. Psychrotrophs A. Grows between 55°C to 65°C

2. Mesophils B. May survive above 60°C

3. Thermophils C. Grow well between 25°C to 45°C

4. Vegetable bacteria D. Grow below 25°C

E. Multiply slowly at 0-40°C

Code

a) 1.B, 2.C, 3.D, 4.A b) 1.C, 2.D, 3.E, 4.A

c) 1.E, 2.D, 3.A, 4.C d) 1.E, 2.A, 3.C, 4.B

176. Match the following growth microorganisms with their respective ranges A to E:

1. Achrommobacter A. Bread spp

2. Aspergillus flavus B. Water supply

3.	Oscillatiria	C.	Meat
4.	Clostridium	D.	Salad
		E.	Milk and cheese products

Code

a) 1.B, 2.C, 3.D, 4.A b) 1.E, 2.A, 3.B, 4.C

c) 1.E, 2.D, 3.A, 4.C d) 1.E, 2.A, 3.C, 4.B

177. Match the following growth microorganisms with their respective appearance of colonies on bismuth Sulphite agar from A to E:

1.	Salemonella typhi	A.	Brown
2.	Salemonell	B.	No growth choleraesuis
3.	Shigella flexneri	C.	Green
4.	Escherichia coli	D.	Yellow
		E.	Black

Code

a) 1.E, 2.C, 3.A, 4.B b) 1.C, 2.D, 3.E, 4.A

c) 1.E, 2.D, 3.A, 4.C d) 1.E, 2.A, 3.C, 4.B

178. Growth curve does not include following phases of bacteria –

a) Decline phase b) Stationary phase

c) Lag phase d) Synchronous growth

179. Bacteria are more sensitive to antibiotics at which phase of growth curve?

a) Decline Phase b) Stationary phase

c) Lag phase d) Long phase

180. The best medium for the production of penicillin is

a) Nutrient agar b) Corn steep liquor

c) Sulfite waste liquor d) Whey

181. Industrial alcohol will be produced by suing starter culture

a) Top yeast b) Middle yeast

c) Bottom d) Feeder yeast

182. A major ingredient of penicillin production media is

a) Can meal b) Corn steep liquor

c) Cane steep liquor d) None of these

183. The outstanding example of traditional microbial fermentation product is

a) Vinegar b) Penicillin

c) Citric acid d) Tetracycline

184. In alcoholic fermentation, CO_2 is evolved during

a) Decarboxylatin of pyruvic acid

b) Formation of acetaldehyde

c) Oxidation of acetaldehyde

d) Both a& b

185. Tobacco and tea leaves are fermented ot give flavor and taste. Theis tpe of fermentation is known as

a) Alcohol fermentation b) Curing

c) Degradation d) Lactic acid fermentation

186. Vinegar fermentation involves

a) Yeasts only

b) Yeasts with lactic bacteria

c) Yeasts with acetic acid bacteria

d) Yeasts with butric acid bacteria

187. Thermo resistant bacteria are important in the preservation of foods by

a) Freezing b) Canning

c) Chemicals d) Inrradiation

188. The fungus used in the industrial production of citric acid:

a) Rhizopus Oryzac b) Fusarium moniliformae

c) Rhizopus nigricans d) Aspergillus nigricans

189. Penicilin is commercially produced by

a) P. notatum b) P.chrysogenum

c) P. citrinum d) P. roqueforti

190. The most commonly used microorganism in alcohol fermentation is

a) A spergilus niger b) Bacillus subtilis

c) Sacharomyces cerevisiae d) Escherichia coli

191. Batch fermentation is also called

a) Closed system b) Open system

c) Fed-Batch system d) Sub-merger system

192. The micro-organism useful for fermentation are

a) Bacteria b) Yeast

c) Fungi d) None of these

193. Over heating of fermentator during fermentation is controlled by

a) Cooling jacket b) Steam

c) Cool air d) None of these

194. Antifoam agent is

a) Silicon compounds b) Corn oil

c) Soyabeen oil d) All of these

195. For the production of ethanol the raw material used is

a) Molasses b) Cellulose

c) Sulphite waste liquor d) None of these

196. Submerged fermentations are

a) Batch fermentation b) Continuous fermentation

c) Both a and b d) None of these

197. Batch fermentation is also called

a) Closed system b) Open system

c) Fed-batch system d) None of these

198. Raw-material used for the production of alcohol is

a) Molasses b) Starch

c) Sulphite waste water d) All of these

199. Microorganisms used for alcohol production

a) Saccharomyces sereviceae b) Bacillus subtilis

c) Penicillium chrysogenum d) None of these

200. For amylase production the microorganism required is

a) B. subtilis b) S. sereviceae

c) A. nigar d) None of these

201. Pecitnase is industrially produced from

a) S. sereviceae b) Trichoderma Koningi

c) A. niger d) None of these

202. Cellulose are produced from

a) S. cerevisiae	b) Trichoderma Koningi
c) A. niger	d) None of these

203. In the production of ethanol industrially the yeast used is

a) K. pneumonia	b) Kluyreromyces fragilis
c) S. sereviceae	d) Bothe b & c

204. Citric acid is used as

a) Flavouring agent is food	b) As an antioxidant
c) As preservative	d) All of the above

205. Citric acids is produced in aerobic conditions by the fungi

a) Aspergillus	b) Penicillin
c) Mucor	d) All of these

206. The raw material for citric production is

a) Corn	b) Molasses
c) Starch	d) None of these

207. The required temperature for the production of citric acid is

a) 10°C - 80°C	b) 30°C - 50°C
c) 20°C - 50°C	d) 25°C - 30°C

ANSWER KEY

1.	**2.**	**3.**	**4.**	**5.**	**6.**	**7.**	**8.**	**9.**	**10.**
a	c	d	c	a	b	b	c	c	b
11.	**12.**	**13.**	**14.**	**15.**	**16.**	**17.**	**18.**	**19.**	**20.**
a	b	b	a	b	d	b	b	c	c
21.	**22.**	**23.**	**24.**	**25.**	**26.**	**27.**	**28.**	**29.**	**30.**
c	b	b	a	d	b	a	b	b	b
31.	**32.**	**33.**	**34.**	**35.**	**36.**	**37.**	**38.**	**39.**	**40.**
b	d	c	c	d	a	c	d	b	c
41.	**42.**	**43.**	**44.**	**45.**	**46.**	**47.**	**48.**	**49.**	**50.**
d	a	d	c	a	a	d	b	d	d
51.	**52.**	**53.**	**54.**	**55.**	**56.**	**57.**	**58.**	**59.**	**60.**
b	d	c	d	c	c	a	c	c	c
61.	**62.**	**63.**	**64.**	**65.**	**66.**	**67.**	**68.**	**69.**	**70.**
c	b	d	a	b	a	b	b	a	c

71.	72.	73.	74.	75.	76.	77.	78.	79.	80.
a	a	b	b	b	b	a	b	b	b
81.	**82.**	**83.**	**84.**	**85.**	**86.**	**87.**	**88.**	**89.**	**90.**
d	c	d	c	b	b	b	a	c	d
91.	**92.**	**93.**	**94.**	**95.**	**96.**	**97.**	**98.**	**99.**	**100.**
b	a	c	e	a	b	b	b	d	a
101.	**102.**	**103.**	**104.**	**105.**	**106.**	**107.**	**108.**	**109.**	**110.**
b	b	b	c	a	c	a	a	d	c
111.	**112.**	**113.**	**114.**	**115.**	**116.**	**117.**	**118.**	**119.**	**120.**
d	c	c	b	d	b	c	a	a	b
121.	**122.**	**123.**	**124.**	**125.**	**126.**	**127.**	**128.**	**129.**	**130.**
d	c	d	b	a	c	b	b	a	c
131.	**132.**	**133.**	**134.**	**135.**	**136.**	**137.**	**138.**	**139.**	**140.**
a	c	b	a	b	b	b	d	c	a
141.	**142.**	**143.**	**144.**	**145.**	**146.**	**147.**	**148.**	**149.**	**150.**
b	b	d	c	c	c	a	b	c	d
151.	**152.**	**153.**	**154.**	**155.**	**156.**	**157.**	**158.**	**159.**	**160.**
a	b	b	d	b	b	d	c	a	d
161.	**162.**	**163.**	**164.**	**165.**	**166.**	**167.**	**168.**	**169.**	**170.**
c	b	a	a	b	a	c	c	b	a
171.	**172.**	**173.**	**174.**	**175.**	**176.**	**177.**	**178.**	**179.**	**180.**
c	b	d	b	a	b	a	d	d	b
181.	**182.**	**183.**	**184.**	**185.**	**186.**	**187.**	**188.**	**189.**	**190.**
c	b	a	d	b	c	b	d	b	a
191.	**192.**	**193.**	**194.**	**195.**	**196.**	**197.**	**198.**	**199.**	**200.**
a	b	a	d	c	c	a	d	a	b
201.	**202.**	**203.**	**204.**	**205.**	**206.**	**207.**			
c	b	d	d	d	a	d			

5

Food Microbiology-IV

1. The method to preserve foods in sealed glass bottles by heat in boiling water was developed by one among the following scientist?

   a) Edward Jenner (1798) b) Francois Nicolas Appert (1804)

   c) Lazzaro spallanzani (1765) d) None of the above

2. Which one of the following person was the first to study Thermophilic bacteria?

   a) A.A Gartner b) Marie von Ermengem

   c) Miquel d) None of the above

3. Clostridium Botulinum is a rod-shaped, Gram-positive bacterium that produces several exotoxins, was first discovered by which one of the following scientist?

   a) Marie von Ermengem b) S.C.Prescott

   c) W. Underwood d) None of the above

4. The sugar fungus, Saccharomyces is organism involved in sugar fermentation, was named in 1837 by which one of the following scientists ?

   a) C. J. Person b) Louis Pasteur

   c) Martinus Beijerinck d) Theodor Schwann

5. Which one of the following method is employed to destroy spore?

   a) Autoclaving b) Fumigation

   c) Gaseous decontamination d) Tyndallization

6. Which one of the following bacterial group produces mycelia structure like fungi ?

   a) Actinomyces b) Proteobacteria

   c) Spirilla d) Vibrio

7. The outer protein shell of a virus is known as which one of the following ?

   a) Bacteriophage b) Capsid

   c) Genome d) Surface receptor

8. In chemical composition of Gram-positive wall, which of the following is presented, except ?

   a) Lipotechoic acid b) Peptodogllycan

   c) Teichoic acid d) All of the above

9. For lab scale purpose, which of the following capacity of fermentor is used ?

   a) 2 lit b) 5 lit

   c) 20 lit d) 100 lit

10. Sauerkraut is fermented by which of the following bacteria, except ?

    a) Acetobacter b) Lactobacillus

    c) Leuconostoc d) Pendiococcus

11. Which of the following is the process of removal of impurity from fermentation broth?

    a) Decantation b) Centrifugation

    c) Filtration d) All of the above

12. Which of the following is the process of extraction of product from fermentation broth ?

    a) Fractional precipitation b) Chromatography

    c) Chemical derivatization d) All of the above

13. Which of one among the following refers to crude protein mixtures obtained from algae yeast and bacterial source ?

    a) SCP b) MCP

    c) Derived protein d) None of the above

14. At a given pH, which one of the following acid can be formed from acetyl coenzyme A and will have higher bacteriostaic effect ?

    a) Acetic acid b) Citric acid

    c) Maleic acid d) Tartaric acid

15. Which one of the following is a source of Achromobacter microorganisms ?

    a) Bread spp. b) Meat scytonema

    c) Milk and Cheese products d) Water supply

16. Which one of the following microbes is well known for attacked if food with low pH value is readily susceptible ?

    a) Bacteria b) Fungi

    c) Protozoans d) Yeast

17. If the coliform bacteria is examined microbiologically in foods which of the following is used preferably ?
    a) Eosine Methylene blue agar
    b) Mac Conkey broth
    c) Violet Red Blue agar
    d) All of the above

18. Which one of the following microorganisms is a source of meat scytonema ?
    a) Achrommobacter
    b) Aspergillusflavus
    c) Clostridium
    d) Oscillatiria

19. Keeping quality of food is related with which one of the following parameter ?
    a) Moisture
    b) pH
    c) Temperature
    d) Water activity

20. Which one of the following microorganisms is source Bread spp.
    a) Achromobacter
    b) Aspergillus flavus
    c) Clostridium
    d) Oscillatiria

21. Oscillatiria is found in which one of the following sources ?
    a) Bread spp.
    b) Meat scytonema
    c) Milk and cheese products
    d) Water supply

22. Which one of the following food items act as a good buffering agent ?
    a) Fruit juice
    b) Milk
    c) Sauerkraut
    d) Soft drinks

23. The post mortem metabolism is an important reaction in degradation of ATP that's result transient accumulation of which one of the following compound, considered to contribute the sensoric appereciation of "fresh fish" taste.
    a) ADP
    b) ATP
    c) Inosine monophosphate (IMP)
    d) None of the above

24. At which one of the following pH most spoilage bacteria grow ?
    a) Acidic pH
    b) Alkaline pH
    c) Neutral pH
    d) None of the above
    e) Oscillatiria

25. The ratio of vapour pressure of solution and vapour pressure of the solvent is expressed by which one of the following ?
    a) Moisture content
    b) Partial pressure
    c) Pressure gradient
    d) Water activity

26. The ratio of the water vapour pressure of food substrate to the vapour pressure of pure water at the same temperature is known as water activity. It can act as which one of the following ?
    a) A processing factor
    b) An extrinsic factor
    c) An intrinsic factor determining the likelihood of microbial proliferation
    d) All of the above
27. The plate count is used normally for estimation of bacterial population in milk, waste foods, soil and many other materials. Generally use the plating medium consisting of which one of the following in plate count ?
    a) Peptone, glucose, sodium chloride, ager and distilled water
    b) Peptone, yeast extract, glucose, sodium chloride and distilled water
    c) Peptone, yeast extract, glucose, sodium chloride, ager and distilled water
    d) Yeast extract, glucose, sodium chloride, agar and distilled water
28. The water activity (a_w) of pure water is _______
    a) 0 b) 0.5
    c) 0.99 d) 1.0
29. Spirulina belongs to which one of the following family ?
    a) Cynophyceae b) Pheophyceeae
    c) Rhodophyceae d) Xanthophyceae
30. Salt and sugar preserve foods because of which one of the following characteristics ?
    a) Deplete nutrients b) Make them acid
    c) Produce a hypertonic environment d) Produce a hypertonic environment
31. Chocolate and confectionary times are having which of the following water activity (a_w) value ?
    a) 0.2-0.3 b) 0.5-0.6
    c) 0.8-0.9 d) 0.95-0.99
32. For alcohol fermentation, which one of the following enzymes is responsible ?
    a) Ketolase b) Oxidase
    c) Peroxidase d) Zymase

33. Pasteur Effect is an inhibiting effect of oxygen on the fermentation process, due to which one of the following ?
    a) Change from aerobic to anaerobic
    b) Non synthesis of ATP
    c) Providing oxygen to an aerobically respiring structures
    d) Rapid utilization of ATP
34. Which one of the following solution is likely to have lover freezing point and lower vapour pressure ?
    a) Milk b) Orange juice
    c) Pure water d) Salt water
35. The practices of post-processing incubation of canned foods in 1813 were introduced by which one of the following scientists ?
    a) Donkin b) Gamble
    c) Hall d) All of the above
36. The first bacteriological study of canning was made by which one of the following persons.
    a) Russell b) S.C. Prescott
    c) W. Underwood d) None of the above
37. By which one of the following methods water activity (a_w) of a solution can be measured ?
    a) Electrical devices b) Freezing point determination
    c) Manometric point d) All of the above
38. Pasteur made an observation concerning with the fermentation of grape juice due to which of the following, except ?
    a) Bacteria producer acid in grape juice
    b) Yeast can grow in sealed or open flasks of juice
    c) Yeast kills can grow and reproduce in juice
    d) Yeast kills by pasteurization to prevent spoilage of juice
39. The microbial spoilage of refrigerated fresh fish has large similarities with that of fresh meat. Pseudomonas is often dominating in the spoilage flora. Similar to these, which one of the following is another spoilage organism specifically associated with marine fishes?
    a) Alteromonas putrifaciens b) Photobacterium phosphoreum
    c) Shewanella putrifaciens d) Both (b) and (c)

40. Nagler reaction detects which one of the following bacteria?

a) Clostridium botulinum b) Clostridium perfringens

c) Clostridium tetani d) Corynebacterium diphtheria

41. Aerobic microbes like Bacillus and pseudomonads require which one of the following reduction potential (Eh) value for cell activity ?

a) Most negative b) Most positive

c) Neutral d) None of the above

42. Which one of the following scientist was not independently showed that microbes are involved in fermentation of sugar to alcohol ?

a) Cagnaird Latour b) F. Kutzing

c) Louis Pasteur d) Teodor Schwann

43. Which one of the following scientist was discovered the existence of forms of life, which can live only in the absence of free oxygen during his studies on the butyric fermentation ?

a) John Needham b) John Tyndall

c) Lazzaro Spallanzani d) Louis Pasteur

44. The presence of fermentable carbohydrates in food results in acid fementation degradation. This phenomenon is due to activity of microbe and known as Which one of the following?

a) Curing b) Pasteuring

c) Rancing d) Sparing

45. The terms 'aerobic' and 'anaerobic' introduced by Which one of the following scientists to designate life in the presence and absence of oxygen, respectively ?

a) Cagnaird Latour b) F. Kutzing

c) Louis Pasteur d) Theodor Schwann

46. Which one of the following starter culture is used for the industrial production of alcohol ?

a) Top yeast b) Middle yeast

c) Bottom yeast d) Feeder yeast

47. Lactenin and anticoliform factore are the food inhibitors, present in Which one of the following food items?

a) Cranberries b) Egg white

c) Milk d) Orange juice

48. Own one of the following yeast that produces carbon dioxide vigorously and has cells tend to become clustered at the surface in brewing?

a) Bottom yeast
b) Feeder yeast
c) Middle yeast
d) Top yeast

49. In which one of the following preservative process of foods, thermo resistant bacteria are important?

a) Canning
b) Chemicals
c) Freezing
d) Irradiation

50. Cranberries and other berries are source of which one of the following natural food inhibitor ?

a) Benzalkonium chloride
b) Benzoic acid
c) Formic acid
d) Propionic acid

51. Fermentation is a process of energy yield when metabolic intermediate act as an electron donors. So, in which of the following case, acetaldehyde $+CO_2$ is produced in a fermentation process?

a) Ethanol fermenters
b) Lactic acid fermenters
c) Both (a) and (b)
d) None of the above

52. The pyruvate dehydrogenase is multi enzyme complex and link to glycolysis with TCA in living cell. Which one of the following bacterial class most likely to have this enzyme?

a) Aerobic bacteria
b) Facultative anaerobic bacteria
c) Microphilic bacteria
d) Strictly anaerobic bacteria

53. Which of the following is the tightly packed, thick walled hyphae usually resistant to heat and desiccation?

a) Basidai
b) Chlamydia
c) Conidia
d) Sclerotia

54. Streptomycin is an antibiotic organic base $C_{21}H_{39}N_7O_{12}$ that is produced by a soil actinomycete (*Streptomyces griseus*), is active against many bacteria, and was the first drug to be used successful against tuberculosis by gram- negative bacteria but is now chiefly used with other drugs because of its toxic side effects. During industrial production of streptomycin, which one of the following is produced by microbes as secondary metabolite in the medium?

a) Vitamin – B_6
b) Vitamin – B_{12}
c) Vitamin _ C
d) Ethanol

55. Industrial fermentated process, along with principal substrate, additional materials are added for better quality. Tobacco and Tea leaves are fermented to give flavor and taste, is used in a process. This type of process is known as which one of the following?

a) Alcohol fermentation b) Curing

c) Degradation d) Lactic acid fermentation

56. Bacillus strains are important as test organisms for sterility testing. Which one of the following bacillus is used to test gamma radiation sterilization?

a) B. mycoides b) B.pumilus

c) B. stearothermophilus d) B. subtilis

57. For the production of various foods, microbial activity of which one of the following microbes is /are responsible.

a) Archaea and bacteria b) Bacteria and protozoa

c) Fungi and bacteria d) Protozoa and fungi

58. In vinegar, acetic acid is the principal component. which of the following organisms(s) are most suitable for this fermentation process?

a) Yeasts only

b) Yeasts with acetic acid bacteria

c) Yeasts with butyric acid bacteria

d) Yeasts with lactic bacteria

59. Which one of the following is the by-product of acetone-butanol fermentation?

a) Isopropanol b) Penicillin

c) Riboflavin d) None of the above

60. Which one of the following genera of bacteria produces alkaline condition and from ropiness in milk?

a) Aeromonas b) Alkaligenes

c) Alteromonas d) Arthrobacter

61. For the selection of lactobacilli present in cheddar cheese, the pH of the medium is maintained at which one of the following?

a) 4.35 b) 5.35

c) 5.75 d) 6.35

62. Which of the following microbe is used as starter organism in industrial production of acetone butanol by fermentation process?

a) Acetobacter acetobutylicum b) Acinetobacter calcoaceticus

c) Clostridium acetobutylicum d) Spingomonas

63. Which one of the following bacteria uses sulfur compound in food as terminal electron acceptor and form hydrogen sulfide?

a) Dehalococcus b) Desulfotomaculum

c) Sulfitobacterium d) Sulfolobus

64. In which one of the following bacterial group almost all genera are pigmented and discolor to the food?

a) Arthrobacter b) Escherichia

c) Flavobacterium d) Pseudomonas

65. Which one of the following is a by-product of acetone-butanol fermentation ?

a) Isopropanol b) Penicillin

c) Riboflavin d) All of the above

66. CO_2 is evolved during which one of the following in alcoholic fermentation?

a) Decarboxylatin of pyruvic acid b) Formation of acetaldehyde

c) Oxidation of acetaldehyde d) Both (a) and (b)

67. Lactobacillus bulgaricus, L. helviticus and L. lactis are an example of which one of the following fermentative type?

a) Heterofermentative b) Homofermentative

c) Mixed acid fermentative d) None of the above

68. A transgenic is an organism whose genetic material has been altered using genetic engineering techniques. These animals are used for improvement of the quality of which of the following food?

a) Milk b) Meat

c) Eggs d) All of the above

69. Thermo-resistant bacteria produce spores that are difficult to destroy by heat. In this case, which one of the following is most suitable for preservation food?

a) Canning b) Chemicals

c) Freezing d) Irradiation

70. The pyogenic, viridians, lactic and eneterococcus groups have been classified in which of the following genera in point of views of food importance?

a) Lactobacillus b) Leuconostoc

c) Staphylococcus d) Streptococcus

71. For industrial production of citiric acid, which one of process is not recommended due to its high cost of production and low recovery?

a) Continuous culture fermentation b) Koji process

c) Liquid surface culture d) Submerged fermentation

72. In batch system of culture, which one of the following organism is used for industrial production of citric acid?

a) Aspergillus nigricans b) Fusarium moniliformae

c) Rhizopus ngricans d) Rhizopus oryzac

73. Acetic acid bacteria (AAB) are important in food microbiology due to which one of the following criteria?

a) Ability to oxidize ethnol; b) Strong oxidizing power

c) To form excess sliminess d) All of the above

74. Which one of the following is the outstanding example of traditional microbial fermentation product?

a) Citric acid b) Penicillin

c) Tetracycline d) Vinegar

75. Which one of the following bacterial genera group Lance field grouping system has been assigned?

a) Lactobacillus b) Leuconostoc

c) Staphylococcus d) Streptococcus

76. Vaccination is the enjecting weak, live, killed or inactivated forms of viruses or their toxins into the person for immunization. First genetically engineered produced vaccine was against which one of the following virus?

a) AIDS b) Hepatitis B

c) Herpes simplex d) Small pox

77. Microbes that can survive to heat treatment like pasteurization are called as which one of the following?

a) Psychrotroph b) Thermoduric

c) Thermoresistant d) Thermosensitive

78. Enzyme immobilization is the process of getting enzyme attached to inert support for its best in industrial purpose. Which method is/are used to get immobilized enzymes?

    a) Adsorption b) Covalent bonding
    c) Encapsulation d) All of the above

79. Vinegar consists mainly of acetic acid and water. Vinegar fermentation involves which one of the following?

    a) Yeast only
    b) Yeast with lactic acid bacteria (LAB)
    c) Yeast with acetic acid bacteria (AAB)
    d) Yeast with butyric acid bacteria (BAB)

80. Which one of the following Raw-material in mostly used for large scale production and cost effectiveness of alcohol?

    a) Cellulose b) Molasses
    c) Starch d) Sulphite waste water

81. Among all which of the following microbe are preferred mostly for alcohol production?

    a) Bacillicus subtilis b) Penicillium chrysogenum
    c) Saccharomyces cervisae d) Schezosaccharomyces pombe

82. Commercially, which one of the following micro-organism is used to produce penicillin?

    a) P. chrysogenum b) P. citrinum
    c) P. notatum d) P. roquefortii

83. Which of the following microbe is/are required for streptomycin production?

    a) Streptomyces griseus b) Streptomyces niger
    c) Saccharomyces cereviceae d) All of the above

84. Which one of the following fermentation process is mostly used for industrial production of acetic acid?

    a) Orleans process b) Rapid process
    c) Submerged process d) All of the above

85. Amylase is an enzyme that cleaves a, 1-4 linkage glycosidic bond and very useful in food industry. Which one of the following microorganism is used for production of amylase?

    a) S. cereviceae b) A. niger
    c) B. subtilis d) None of the above

86. Escherichia coli is the sole member of the bacterial genus Escherichia and most widely used in food microbiology in an addition of experimental cell system in biochemistry and molecular biology. E. coli produce which type of toxins?

a) Exotoxins b) Endotoxins

c) Leucocidin d) Both (a) and (b)

87. Pectinase has been used in food industry fro degrading pectic substances in raw fruit and vegetables. Which of the following micro-organism is used for industrial production of pectinase?

a) Aspergillus niger b) S. Cerviceae

c) Trichoderma koningi d) None of the above

88. Cellulose is the main dietary source and bulking agent in food industry. Which of the following microbe is used widely for cellulose production?

a) Acetobacter b) S. cerviceae

c) Trichoderma koningi d) None of the above

89. Which one of the following microbe plays an important role in the production of vitamin B_{12} by using glucose as Carbon source?

a) Propionibacterium spp. b) Pseudomonas spp.

c) Both (a) and (b) d) None of the above.

90. Clostridia have wide application for its fermentative use of substrate while universally well-known is the group of anaerobic bacilli mostly occurs in. Clostridium acetobutylicum is used fro the production of which one of the following product?

a) Acetone- Butanol b) Ethanol

c) Vitamin B_{12} d) None of the above

91. Acetic acid bacteria (AAB) are capable of acitification and use energy from the oxidation of ethanol to acetic acid during fermentation process. which one of the following is an acetic acid bacterium ?

a) Acetobacter spp. b) Acetobacterium woodii

c) Gluconobacter spp. d) Both (a) and (c)

92. Citric acid is an important metabolite produced in Kreb's cycle as first stable product. In fermentation process, which one of the following fungi is used to produce citric acid during aerobic conditions ?

a) Aspergillus b) Mucor

c) Penicillin d) None of the above

93. Which one of the following is used as raw material for citric acid production?

a) Corn b) Molasses

c) Starch d) None of above

94. Aspergillus niger is generally used for the production of which of the following compound?

a) Citric acid b) Ethanol

c) Lactic acid d) penicillin

95. At what pH among the following Citric acid production is maximum during fermentation process in the fermenter ?

a) 7.0 b) 5.0 to 6.0

c) 8.0 to 9.0 d) to 6.0

96. What is the temperature require for the production o citric acid ?

a) 10°C to 80°C b) 20°C to 70°C

c) 30°C to 50°C d) 50°C to 70°C

97. Which one of the following strain of fungi is used for the large scale penicillin production ?

a) Penicillium chrysogenum b) P. notatum

c) Saccharomyces spp. d) Streptomyces aurecus

98. Which one of the following enzyme is playing an important role in preparation of 6-amino penicillic acid (APA) from penicillin spp ?

a) Acylase b) Penicillinone

c) Penicillin acylase d) None of the above

99. The different aerobic Colony Counts (ACCs) between food categories reflect to which of the following, except ?

a) Expected level of contamination of the raw material

b) Potential for microbial growth during storage

c) Potential shelf life

d) Hazard analysis and critical control

100. Which one of the following is not true regarding Acetic acid bacterial ?

a) Aectobacterium b) Aerobic

c) Gram-negative d) Rod-shaped

101. Which one of the following is an antifoaming agent, very effective for the reduction of foam produced during fermentation?

a) Corn oil b) Silicon compounds

c) Soyabean oil d) All of the above

102. Fermentors are use in laboratory for small scale production and product evaluation. What is one of the following is the capacity of laboratory fermentors generally used?

a) 10000 gallons b) 12-15 liters

c) 2000 gallons d) 500 liters

103. Penicillin is commercially produced by which one of the following species ?

a) *P. notatum* b) *P. chrysogenum*

c) *P. citrinum* d) *P. roquefortii*

104. Which micro-organism is used most widely for alcohol production in industrial process ?

a) *Aspergilus niger* b) *Bacillus subtilis*

c) *Escherichia coli* d) *Sacharomyces cervisiae*

105. Vitamin B_{12} can be estimated and determined by using which one of the following micro-organism ?

a) *Bacillus subtilis* b) *E. Coli*

c) *Lactobacillus leichmanni* d) *Lactobacillus* spp.

106. Acetification is which one of the following type conversion of ethanol to acetic acid by bacteria (usually *Acetobacter* spp.)

a) Aerobic conversion b) Anaerobic conversion

c) Fermentative conversion d) None of the above

107. Batch fermentation is also known as which one among the following ?

a) Closed system b) Fed-Batch system

c) Open system d) Sub-merger system

108. Streptokinase is producted by which one of the following microbe ?

a) Bacillius b) Salmonella

c) Staphylococcus d) Streptococcus

109. Which one of the following is used as the best medium for the production of Penicillin ?

a) Corn steep liquor b) Nutrient agar

c) Sulphite waste liquor d) Whey

110. Which one of the following micro-organisms is called as brewer yeast ?

a) Saccharomyces boulardii b) Saccharomyces cerevisiae

c) Saccharomyces ludwigi d) Saccharomyces pastorianus

111. Butanol is obtained by fermenting molasses, which one of the following microbe plays their role?

a) Clostridium butyricum and Clostridium acetobutylicum

b) Clostridium butyricum and clostridium tetanai

c) Clostridium butyricum and Lactobacillus

d) Clostridium butyricum and Clostridium oceanicum

112. The microbial decomposition of high protein materials (like meat and eggs) are responsible formation of foul-smelling products is known as which one of the following ?

a) Fouling b) Putrefaction

c) Rancidity d) None of the above

113. Which one of the following is the source of vit.-A in food supplement?

a) Rhodotorula gracilis b) Sterptococcus

c) Yeast d) Both (a) and (b)

114. Yeast cells are good source of which of the following vitamin ?

a) Vit. A and vit. B b) Vit. A and vit. D

c) Vit. B and vit. D d) All of the above

115. Roquefort cheese is ripened by which of the industrially important Penicillium ?

a) Penicillium comemberti b) Penicillium rhizogenes

c) Penicillium roqueforti d) None of the above

116. Penicillium camemberti is used for ripeing of which of the following cheese?

a) Roqueforti cheese b) Camembert cheese

c) Both (a) and (b) d) None of the above

117. Unicelled microbes grown as source of proteins are known as which one of the following proteins ?

a) Microbial proteins b) Single cell proteins

c) Unicelled proteins d) None of the above

118. Which one of the following is the rich source of protein and used as single cell protein ?
   a) Chlorella and Scendesmus b) Scenedesmus
   c) Spirulina and Chlorella d) All of the above

119. Zymase is an enzyme used for cleaving of glycosidic bond in starch solution during fermentation. Industrial production of this enzyme is obtained from which one of the following microbe ?
   a) Saccharomyces boulardii b) Saccharomyces cerevisiae
   c) Saccharomyces ludwigi d) Saccharomyces ludwigi

120. Glucose oxidase is used for measuring free glucose in the body as measure of disease diagnosis. From one of the following microbe. Glucose oxidase is obtained ?
   a) Aspergillus niger b) Penicillium spp.
   c) Saccharomyces cerevisiae d) Spirulina

121. Bacillus thuringiensis is a soil microbe that produce crystal protein toxin against many vector of diseases. It is mainly used as which one of the following ?
   a) Fungicide b) Insecticide
   c) Microbicidal agent d) Rodenticide

122. In most oil spills and petroleum sludge contamination, which of the following is most fount Petroleum degrading species ?
   a) Micrococcus b) Nocardia
   c) Penicillium d) All of the above

123. Kojic acid acts as an important component for skin revival by removing free oxygen and phenolic compounds. It is obtained from which one of the following microbe ?
   a) Aspergillus b) Micrococcus
   c) Nocardia d) Penicillium

124. Fumaric acid is obtained from which of the following fungus ?
   a) Aspergillus niger b) Penicillium spp.
   c) Rhizopus stoloniger d) Saccharomyces cerevisiae

125. In alcohol fermentation, which of the following is the most commonly used microorganism?
   a) Aspergilus niger b) Bacillus subtilis
   c) Escherichia coli d) Sacharomyces cerevisiae

126. Which of the following is the basic principle in industrial microbiology ?

a) Suitable growth conditions b) Fermentation
c) Providing aseptic conditions d) All of the above

127. Which one of the following is the large vessel among all, containing all the parts and necessary condition for the growth of desired micro-organisms?

a) Bio reactor b) Auto reactor
c) Impeller d) None of the above

128. Salmonellosis is caused by salmonella pathogen. It can be passed by which one of the following ?

a) Pasteurized food b) Turtles
c) Unpasteurized cheese d) None of the above

129. Giardiasis caused by which one of the following microorganism in human ?

a) Bacteria b) Protozoans
c) Virus d) None of the above

130. Which one of the following can be used for inhibiting bacterial growth in food preservation ?

a) Dehydration b) High salt and High sugar
c) Low pH and High salt d) All of the above

131. Which of the following food materials is considered as most likely to be contaminated with mercury ?

a) Beef b) Poultry
c) Salmon d) Vegetable

132. Cryptosporidiosis is the disease caused by cryptosporidia-parasite in intestine of human. Which one of the following is reason for cryptosporidiosis ?

a) Blood b) Contaminated water
c) Food d) None of the above

133. Which one of the following is the easiest way to prevent food borne illness?

a) Avoiding spoiled food b) Hand washing
c) Milk and milk products d) Raw food avoiding

134. Which of the following micro-organism cannot be killed by heat alone in intoxicated food ?

a) Bacterial spore b) Parasites
c) Prions d) Viruses

135. The contaminated food acts as mean of disease and sickness due to bacterial pathogenesis. Which one of the following is a food borne illness caused by a pathogen ?

a) Gastro intestinal b) Infection

c) Parasites d) Poisoning

136. Exotoxin is heat labile toxin produced by bacteria. Which of the following statement is incorrect regarding the bacterial exotoxin ?

a) Activity secreted by bacterial cell

b) Are components of lipo-polysaccharides

c) Produced by both gram + ve and – ve

d) They are proteins

137. Citric acid is recommended for use as which of the following in food industry, except?

a) Antioxidant b) Coloring agent

c) Flavouring agent d) Preservative

138. Size of fermenter depends on which one of the parameters ?

a) Total capacity b) Total media

c) Total volume d) All of the above

139. Type construction medium used to construct fermenter depends on which one of the following ?

a) Type of fermentation b) Type of media

c) Type of nutrients d) Type of organism

140. Which one of the following part is used for vigorous stirring and agitation of media?

a) Impeller b) Motor

c) Sparger d) Spindle

141. Fermentation is carried out in the presence of oxygen is called as which one of the following ?

a) Aeration b) Assimilation

c) Fermentation d) Respiration

142. Foam production is more in medium, if containing which one of the following biomolecules ?

a) Amino acid b) Carbohydrate

c) Lipid & Fats d) Protein & Peptides

143. Foam is controlled in fermentation by the application of which of the following agent ?

a) Biological agent
b) Chemical agent
c) Physical agent
d) All of the above

144. Which one of the following is used as antifoaming agent ot reduce foam during product recovery in fermentation tank ?

a) Media
b) Nutrient
c) Oils
d) None of the above

145. Growth medium is used for the cultivation of microorganism is known as which one of the following ?

a) Fermentation media
b) Growth media
c) Nutrient media
d) None of the above

146. C. N. and O are the principal component for microbial growth. Instead of which one of following source, glucose is used as in fermentation ?

a) Carbon
b) Hydrogen
c) Nitrogen
d) Sulphur

147. Depending on the composition, fermentation media categorized as which one of the following category ?

a) Complete and Simple Media
b) Complex Media
c) Composition and Compost Media
d) Simple and Complex Media

148. Using continuous culture system, microbial population is maintained by which one of the following phase for long time ?

a) Death
b) Exponential
c) Lag
d) Log

149. A chemostat is also known as which one of the following?

a) Bacterial filter
b) Bacteriogen
c) Bacteriogen vessel
d) Bactogen

150. Cell density is controlled in which one of the following culture conditions by increasing and decreasing flow of culture medium ?

a) Chemostat
b) Continuous culture
c) Synchronous culture
d) Turbidostat

151. Resist sudden change is pH in which one of the following combination of monobasic and dibasic slat ?

a) Acid b) Ankali

c) Buffer d) None of the above

152. Enrichment is done usually to isolate indigenous microbe from their habitat. Which one of the following is best describing for an enrichment culture ?

a) An infectious culture

b) Something that inhibits growth for all microorganisms

c) Something that provides growth for a certain microorganism but not for others

d) Something that provides growth for all microorganisms.

153. Oxygen acts as terminal electron acceptor in respiration process. Microorganisms require oxygen are known as which one of the following?

a) Facultative anaerobes b) Obligate aerobes

c) Obligate anaerobes d) None of the above

154. Which one of the following process allows enrichment and concentration in one step, by reducing the volume of material for further processing?

a) Diffusion b) Filtration

c) Precipitation d) All of the above

155. The fermentation process is carried in batches is called as which one of the following ?

a) Batch fermentation b) Continuous fermentation

c) SSF d) Submerge fermentation

156. Fermentation is carried out in a glass coated vessel is called as which one of the following?

a) Baffles b) Chemostat

c) Fermenter d) Turbidostat

157. In the Industrial Microbiology, which one of the following is considered as the basic principle ?

a) Fermentation b) Providing aseptic conditions

c) Suitable growth conditions d) All of the above

158. Microbial growth under industrial fermentation conditions usually utilizes which one of the following metabolism of the organism ?

a) Luxury b) Rapid

c) Slow d) Co-metabolism

159. The Crowded- Plate technique is used to produce which one of the following product by screening of bioactive compound from microbes ?

a) Antibiotic b) Growth factor

c) Vitamin d) Both (a) and (b)

160. For amino acid production, which one of the following microorganisms is used for strain improvement by metabolic engineering ?

a) Aspergillus b) Corynebacterium

c) Escherichia d) Rhizobium

161. Which one of the following is used for through mixing of medium and inoculum in fermentor ?

a) Headspace b) Impeller

c) Shaft d) Sparger

162. Which one of the following portions in the fermentor is used for evaluation of medium addition and product formation?

a) Head space b) Impeller

c) Shaft d) Sparger

163. During the fermentation, overheating of fermentator is controlled by which one among the following part of fementator ?

a) Cool Air b) Cooling jacket

c) Steam d) None of the above

164. Which one of the following characters is not the target of strain improvement?

a) Genetic stability b) Slow growth

c) Non toxicity d) Genes for enzymes

165. Which one of the following is an optimization process is/are required for strain improvement?

a) Carbon source b) Nitrogen source

c) Precursor d) All of the above

166. Which one of the following microbe is recognized as GRAS status in industrial microbiology ?

a) Bacillus subtilis b) Escherichia

c) Mucor d) Sphingomonas

167. For stain improvement, which one of the following Fungi was first reported to be used ?

a) Aspergillus b) Penicillum

c) Rhizoctonia d) Yeast

168. Which one of the following compounds is used in the precipitation of proteins mainly due to changes in the dielectric constant ?

a) Ethanol, Ethane b) Acetone, Ketone

c) Glycol, Glycerol d) Ethanol, Acetone

169. Which one of the following processes is used at all scales of operation to separate suspended particles from a liquid or gas, using a porous medium which retains the particles but allows the liquid and gas to pass ?

a) Adhesion b) Diffusion

c) Filtration d) Precipitation

170. Which type of filter is suitable for "polishing" large volumes of liquid with low solid content or small batch filtrations of valuable solids ?

a) Cross flow filter b) Diffusion filter

c) Frame filter d) Pressure leaf filter

171. Cross flow filtration is also known as which one of the following ?

a) Continuous filter b) Rotary filter

c) Stacked filter d) Tangential filtration

172. An illness is considered when cause infection in body by which one of the following ?

a) Campylobacter b) Clostridium

c) Escherichia d) Salmonella

173. Which one of the following proteobacteria is known as acetic acid producers?

a) Acetobacter b) Azospirillum

c) Gluconobacter d) Both (a) and (c)

174. Radappertization is a method that uses gamma rays from which one of the following source for control of microorganisms is foods ?

a) Magnesium (Mg) b) Cobalt (Co)

c) Cupper (Cu) d) Iron (Fe)

175. Which one of the following microbes associated with an optimum growth temperature of 37° C and utilizes glucose with the production of acid and gas ?

a) Campylobacter jejuni b) Escherichia coli
c) Salmonella spp. d) Shigella spp.

176. Which of the following bacteria are common contaminants in fruits and vegetables, except ?

a) Bifidobacterium b) Corynebacterium
c) Erwinia d) Pseudomonas

177. Which of the following bacteria are common contaminants in fresh meat, except?

a) Acinetobacter b) Aeromonas
c) Pseudomonas d) Brevundimonas

178. Which of the following bacteria are common contaminants in milk, except ?

a) Lactobacillus b) Lactococcus
c) Leuconostoc d) Pseudomonas

179. In milk, common contaminants bacteria is which of the following, except ?

a) Erwinia b) Proteus
c) Pseudomonas d) Streptococcus

180. Which one of the following fungi are not common contaminants in milk ?

a) Phytophthora b) Rhizopus
c) Both (a) and (b) d) None of the above

181. The fungi are common contaminants in milk, except which one of the following ?

a) Alternaria b) Aspergillus
c) Botrytis d) Mucor

182. Which of the following are known as common contaminants fungi in milk, except ?

a) Cladosporium b) Geotrichium
c) Penicillium d) Rhizopus

183. Which of the following bacteria are common contaminants in high- sugar foods, except ?

a) Bacillus b) Clostridium
c) Flavobacterium d) None of the above

184. In high-sugar foods, which of the following fungi are common contaminants, except ?

a) Penicillium
b) Saccharomyces
c) Torulla
d) None of the above

185. Mycotoxins are produced by which of the following fungi in foods, except ?

a) Asperigillus flavus
b) Aspergillus ochraceus
c) Aspergillus parasiticus
d) Penicillium viridicatum

186. During storage, which one of the following will minimize deterioration of fats and oils ?

a) Anticaking agents
b) Antioxidants
c) Emulsifiers
d) Preservatives

187. Nitrite is often used in canned meat products to inhibit the growth of which one of the following microorganisms ?

a) Salmonella spp.
b) Bacillus cereus
c) Clostridium Botulinum
d) Listeria monocytogenes

188. Which one of the following food products is most likely to be associated with food poisoning caused by clostridium Botulinum ?

a) Boiled rice
b) Canned sweet corn
c) Canned tomatoes
d) Roast chicken

189. Match the following organisms with their characteristics properties that dominate the spoilage flora on cold-stored meat in different atmospheres.

1.	Pseudomonas	A.	Very CO_2- resistant, indifferent to oxygen
2.	Enterobacteriaceae and Aeromonas	B.	Relatively CO_2 resistant, resistant to low a_w
3.	Brochothrix thermosphacta	C.	Aerobic, very CO_2 – sensitive
4.	Lactobacillus	D.	Facultative, intermediate CO_2-sensitivity

(a) 1B 2C 3D 4A
(b) 1C 2D 3B 4A
(c) 1C 2D 3A 4B
(d) 1D 2C 3A 4B

190. Match the following Mycotoxins with their sources.

1.	Aflatoxins	A.	Pc. Exapnsum and Pc roqueforti
2.	Ochratoxin A	B.	Pc. Cyclopium & Pc viridicatum

3. Patulin — C. Asp. Ochraceus and Penicillium viridicatum

4. Penicillinic acid — D. Aspergillus flavus and Asp. Parasiticus

5. Zearalenon — E. Fusarium graminearum

a) 1B 2C 3D 4A 5E b) 1C 2D 3B 4A 5E

c) 1C 2D 3A 4E 5B d) 1D 2C 3A 4B 5E

191. Match the following cause with their suitable food born disease.

1. Patulin — A. Liver cancer

2. Zearalenon — B. Kidne/Liver damage, teratogenic

3. Ochratoxin — C. Diarrhoea

4. Aflatoxins — D. Infertility

a) 1B 2C 3D 4A b) 1C 2D 3B 4A

c) 1C 2D 3A 4B d) 1D 2C 3A 4B

192. Match the following bacteria with their shape.

1. Road shape — A. Pasteuria

2. Curved bacteria — B. Bacillus, Lactobacillus

3. Filamentous shaped — C. Spirillum e.g., Aquaspirillum

4. Pear shaped — D. Streptomyces spp.

5. Lobed spheres — E. Sulfolobus

a) 1B 2C 3D 4A 5E b) 1C 2D 3B 4A 5E

c) 1C 2D 3A 4E 5B d) 1D 2C 3A 4B 5E

193. Match the following bacteria with their size.

1. Spirillum parvum — A. 0.8 mm

2. Bacillus megaterium — B. 1 x 3 mm

3. E. coli — C. 0.1 to 0.3 m

4. S. pneumoniae — D. 1.5 x 4 m

5. Haemophilus influenza — E. 0.25 x 1.2 mm

a) 1B 2C 3D 4A 5E b) 1C 2D 3B 4A 5E

c) 1C 2D 3A 4E 5B(d) 1D 2C 3A 4B 5E

194. Which among the following microbes is not a prime concern for deterioration of water quality in drinking water?

a) Legionella b) Shigella

c) Vibrio parahaemolyticus d) Vibrio vulnificus

195. Which of the following method is used for removal of suspended materials from waste water ?

a) Filtration b) Purification
c) Sedimentation d) Settlement

196. Disinfection of water during water purification is usually done with application of which of the following chemical ?

a) Chlorine b) Ozone
c) Both (a) and (b) d) None of the above

197. Most of the spores found in the drinking water belong to which of the following protest ?

a) Cryptosporidium oocysts b) Cyclospora
c) Giardia intestinalis d) All of the above

198. Which one of the following test is used as presumptive test for enumeration of coliform in water samples ?

a) Most probable number (MPN) b) Hetercoliform count (HCC)
c) Aerobic colony count (ACC) d) None of the above

199. Which of the following process total carbon content is measured in waste water ?

a) Total organic carbon (TOC)
b) Chemical oxygen demand (COD)
c) Biochemical oxygen demand (BOD)
d) All of the above

200. Which of the following is the amount of dissolved O_2 needed for microbial oxidation of biodegradable organic matter ?

a) Biochemical oxygen demand (BOD)
b) Chemical oxygen demand (COD)
c) Total organic carbon (TOC)
d) All of the above

201. Trickling filters are used for removal of organic waste from waste water by secondary treatment process. Which of the following system works for trickling filters ?

a) Aerobic b) Anaerobic
c) Microacerophilic d) All of the above

202. Most of the proteinaceous compounds are degraded after digestion of carbohydrate with the help of which of the following bacteria?
   a) Bacteroides b) Butyrivibrio
   c) Clostridium d) All of the above

203. For detection of fecal coliform, which of the following medium is used in completed test of coliform detection ?
   a) EMB ager b) LES endo agar
   c) Brilliant green lactose bile ager d) All of the above

204. Which one of the following sterilization process does not kill bacterial endospores ?
   a) Autoclave b) Dry heat
   c) Filtration d) None of the above

205. HPLC has several advantages. Which one of the following are not true statements about HPLC?
   a) Adaptable to very small sample sizes or large-scale, preparative procedures
   b) Columns can be reused without repacking or regeneration
   c) May be developed in ascending or descending solvent flow
   d) Resolution and speed of analysis far exceed the classical methods

206. In which of the following techniques, the stationary phase consists of inert particles that contain small pores of a controlled sized ?
   a) Partition chromatography b) Gel filtration chromatography
   c) Gel electrophoresis d) None of the above

207. Heat labile solutions like enzyme, hormone and vitamis can be sterilized by which one of the following sterilization process ?
   a) Autoclave b) Membrane filtration
   c) Dry heat method d) Pasteurization

208. Which of the following is the killing, inhibition, or removal of microorganisms that may cause disease ?
   a) Sterilization b) Disinfection
   c) Sanitization d) None of the above

209. Which one of the following scientist is the inventor of microscope ?
   a) Antony van b) Galileo
   c) Koch d) Louis Pasteur

210. Southern blotting refers to which one of the following ?
   a) Comparison of DNA fragments from two sources
   b) Transfer of DNA fragments from electrophoresis gel to nitrocellulose sheet
   c) Attachment of probes to DNA fragments
   d) Transfer of DNA fragments from in vitro cellulose membrane to electrophoresis gel

211. Which one of the following is most resistant to antimicrobial agent ?
   a) Endospore b) Mature cell
   c) Vegetative cell d) Younger cell

212. If effective mass of the sedimenting particle, angular velocity of rotation in rad/sec and distance of the migrating particles from the central axis of rotation are indicted by, m, w and r, respectively. The Intensity of the centrifugal force (F) is defined by which one of the following equation ?
   a) $F = mw^2r$ b) $F = mw^2/r$
   c) $F = mw^2r/r$ d) None of the above

213. In time-temperature combination, on the basis of which one of the following for HTST pasteurization of 71.1 °C for 15 sec is selected ?
   a) Bacillus subtilis b) Clostridium Botulinum
   c) Coxiella burnetii d) Escherichia coli

214. Which one of the following conditions is maintained in autoclave during sterilization?
   a) 120 °C and 25 pounds of pressure
   b) 121 °C and 15 pounds of pressure
   c) 131 °C and 15 pounds of pressure
   d) 131 °C and 25 pounds of pressure

215. Which one of the following is used as an indicator microbe for detection of proper sterilization ?
   a) Geobacillus stearothermophilus
   b) Thermococcus acidocaldareus
   c) Herminiimonas arsenicoxydans
   d) Bacillus thermophilus

216. Which of the following can be sterilized by pasteurization, except ?
   a) Milk b) Orange juice
   c) Beer d) Vitamins

217. Which one of the following is referred as the lowest temperature at which a microbial suspension is killed in 10 minutes ?

a) Decimal reduction time (D) b) Thermal death point (TDP)

c) Thermal death time (TDT) d) None of the above

218. In chromatography techniques, which one of the following is common in both mobile and stationary phase ?

a) Gas b) Liquid

c) Solid d) None of the above

219. Aerobic plate count (APC) is also known as which one of the following, except ?

a) Aerobic colony count (ACC) b) Standard plate count (SPC)

c) Total viable count (TVC) d) Most probable number count (MPNC)

220. Who one among the following was introduced Petri dish as a suitable medium container for the culture of bacteria?

a) Joseph Lister b) Richard. J. Petri

c) Francois Appert d) None of the above

221. The Shortest time is needed to kill all organisms in a microbial suspension at a specific temperature and under defined conditions is known as which one of the following ?

a) Decimal reduction time (DRT) b) Thermal death point (TDP)

c) Thermal death time (TDT) d) None of the above

222. Decimal reduction time (D) or D value is associated with which one of the following?

a) Time in minutes at a specific temperature (usually 250 °F or 121.1°C) needed to kill a population of cells

b) Time needed to kill all organisms in a microbial suspension at a specific temperature

c) Time needed to kill all organisms in a microbial suspension at given temperature

d) Time required to kill 90% of the microorganisms or spores in a sample at a specified temperature

223. High-efficiency particulate air (HEPA) filters are used in laminar hoods to remove the particles of which one of the following size ?

a) 0.1 μm b) 0.3 μm

c) 0.5 μm d) 0.25 μm

224. For killing microbial pathogens, which one of the following are the two most popular alcohol germicides ?

a) Ethanol an isopropanol
b) Ethanol and propanol
c) Methanol and ethanol
d) Methanol and propanol

225. The separation of molecules in Chromatography is based on which one of the following between the mobile and stationary phases ?

a) Differential partitioning
b) Partition coefficient
c) Sedimentation coefficient
d) None of the above

226. After drying , which one of the following bacteria dies very quickly ?

a) Mycobacterium tuberculosis
b) Pseudomonas aeruginosa
c) Staphylococcus aureus
d) Terponema palladium

227. In Gram Staining method, the fixed bacterial smear is subjected to the following staining regents in which one of the following order of sequence ?

a) Crystal violet → Iodine solution → Alcohol (decolorizing agent)→ Safranin
b) Crystal violet → Iodine solution → Safranin → Alcohol (decolorizing agent)
c) Crystal violet → Safranin → Alcohol (decolorizing agent) → Iodine solution
d) Iodine solution → Crystal violet → Alcohol (decolorizing agent) → Safranin

228. Which of the following stains is used frequently to identify mycobacterium and other bacteria whose cell walls contain high amount of lipid.

a) Gram stain
b) Negative stain
c) Acid fast stain
d) Spore stain

229. Which of the following stains is used to classify microorganisms based on their cell wall content?

a) Gram stain
b) Negative Stain
c) Acid fast stain
d) Methylene blue

230. The order of reagent used in Gram staining are

a) Crystal violet, iodine, safranin, alcohol
b) Crystal violet, iodine, alcohol, safranin
c) Crystal violet, safrani, alcohol, iodine
d) Iodine, crystal violet, safranin, alcohol

231. Capsules of bacteria are viewed by
   a) Spore staining b) Scanning EM
   c) Gram staining d) ZN staining
   e) Simple stain

232. Structure essential for survival of most bacteria
   a) Cell wall plasma membrane b) Plasma membrane
   c) Capsule d) a & b
   e) a,b,c

233. Freezing contaminated eat would most likely kill
   a) Bacteria b) viruses
   c) Worm parasite d) Bacterial spore

234. If a disinfectant achieves a 90% kill of a microbial population in 1 min, what is he minimum time required to achieve a 100% kill of 10,000 microbial cells in a test solution?
   a) 3 min b) 6 min
   c) 5 min d) 10 min

235. UV light is less effective in killing bacteria because
   a) Cell wall of bacteria absorbs UV light and protected
   b) Bacteria have DNA repair mechanism
   c) The cell membrane of bacteria absorb the UV light
   d) Bacteria reflect rather than adsorb UV light

236. Gamma ray and X rays are effective in killing microorganism because they
   a) Dislodge electrons from atoms, creating ions
   b) Damage DNA
   c) Produce powerful oxidizing agents
   d) All of above

237. The active antimicrobial ingredient in bleach is
   a) Phenol b) Hydrochloride
   c) Hypochloride d) Iodine

238. The minimum time used for sterilization by autoclaving
   a) 5 min b) 15 min
   c) 45 min d) 1 hour

239. Which of the following is a limitation of the autoclave ?
   a) Length of time b) Ability to inactivate virus
   c) Ability to kill endospores d) Use with heat sensitive materials

240. Botulinum toxin is an example of a (n)
   a) Endotoxin b) Haemolysis
   c) Lip polysaccharide d) Exotoxin

241. If a bacterium cell having generation time of 30 minutes, is placed in a suitable broth at time 0, what will be the cell numbers after 4 hours of incubation
   a) 256 b) 1285
   c) 64 d) 96

242. The collapse of a cell due to water loss is called
   a) Hydrolysis b) Halophile
   c) Osmoregulation d) Plasmolysis

243. Bacteria that live in high slat concentration are called
   a) Halophiles b) Acidophiles
   c) Mesophiles d) Alkaliphies

244. Which type of bacteria are also called blue green algae?
   a) Eubacteria b) Parmecium
   c) Protists d) Cyanobacteria

245. Which of the following can not be treated with anti-microbial drugs ?
   a) Atypical pneumonia b) Q fever
   c) Common cold d) T.B.

246. Claviceps purpurea, a major food contaminant is a
   a) Fungus b) Protozoa
   c) Algae d) Bacterium

247. Which of the following found on fruits and vegetables is least likely to cause disease in humans ?
   a) Salmonella sp. b) Shigella sp.
   c) Pseudomonas fluorescens d) Ascaris sp.

248. A green discoloration of refrigerated meat may be caused by the growth of
   a) Pseudomonas syringae b) Monilia sitophila
   c) Rhizopus mingricans d) Pseudomonas meephitica

249. Approximately 50 % of the poultry food based outbreaks in restaurants are caused by
   a) *Salmonella* sp. b) *Clostridium perfringens*
   c) *Staphylococcus aureus* d) *E.coli*
250. Hens may lay infected eggs if they are infected with
   a) *Clostridium prefringens* b) *Staphylococcus aureus*
   c) *E. coli* d) *Salmonella pullorum*
251. Which component of egg white helps to kill bacteria that may invade the egg
   a) Lysozyme b) H_2O_2
   c) HCI d) Chlorine dioxide
252. The most common microorganism in freshly drawn milk is
   a) *Staphylococcus epidermidis* b) *Staphylococcus aureus*
   c) *M. bovis* d) None of these
253. A pathogen that can grow on milk is
   a) *Coxiella burnetti* b) *M. bovis*
   c) *Brucella* sp. d) *L. monocytogens*
   e) All of these
254. Useful products obtained from microbes are
   a) Amino acids b) Alcohol
   c) Single cell protein d) All of these
255. Food processed by commercial may contain endospores of
   a) *Monilia sitophila* b) *Rhiszopus nigricans*
   c) *Bacillus stearothermophilus* d) *Clostridium perfringens*
256. Aflatoxins are produced by a
   a) Bacterium b) Virus
   c) Fungus d) Protozoa
257. Which of the following virus requires a surrogate virus infection in order to cause disease ?
   a) Hepatitis B b) Hepatitis d
   c) Norwalk virus d) Rotavirus
258. Bacterial cells are metabolically uniform during
   a) Lag phase b) Log phase
   c) Transition from lag to log phase d) Transition from log to stationary phase

259. Total bacterial count is

a) Total living bacteria in a simple

b) Total living and dead bacteria in a sample

c) Total bacteria in long phase

d) None

260. The viable count of bacteria is maximum during

a) Lag phase b) Log phase

c) Stationary phase d) Phase of decline

261. The following enzymes may destroy penicillin

a) β lactamase b) Amidase

c) Both d) None

262. Holoenzyme is

a) Apoenzyme + cofactor b) Apoenzyme + coenzyme

c) Co-enzyme + cofactor d) None

263. Fats and oil can be sterilized by

a) Autoclave b) Not air oven

c) Tyndalization d) None of the above

264. Cold sterilization is

a) Sterilization at low temperature

b) Sterilization by chemical agent

c) Sterilization by ionizing radiation (γ ray)

d) Sterilization by HCHO or ethylene oxide

265. Gelatin melts at

a) 20° C b) 25° C

c) 35° C d) 45° C

266. Agar melts at

a) 98° C b) 80° C

c) 70° C d) 60° C

267. Agar solidifies at

a) 45° C b) 25° C

c) 30° C d) 35° C

268. Agar is composed of

a) Protein b) Lipid

c) Carbohydrate d) All of the above

269. The source of Agar is

a) Pond water weeds b) Agriculture bye product

c) Marine algae d) None of the above

270. The isolation of bacteria in pure culture by dilution method was first used by

a) R. Koch b) F. Hesse

c) Lister d) None of the above

271. Bacteria that grow at 50-55° C are called

a) Thermophiles b) Psychrophiles

c) Mesophiles d) Halophiles

272. Organism that can grow only in the presence of O_2 are called

a) Facultative anaerobes b) Obliate anaerobes

c) Obligate aerobes d) Microaerophilic

273. In anaerobes

a) O_2 is toxic b) O_2 is not toxic

c) Caltalase is present d) None of these

274. Bacteria that can grow in the presence or absence of air are known as

a) Aerobes b) Anaerobes

c) Facultative anaerobes d) Microaerophilic

275. Which of the following diseases are transmitted through milk?

a) Q fever b) Brucellosis

c) Scarlet fever d) T.B.

e) All of the above

276. Which of the following do not cause food poisoning ?

a) B. cereus b) Cl. Botulinum

c) Streptococcus pyogenes d) Salmonella

277. Organisms usually found in silage

a) *Leptospira* b) *B. subtilis*

c) *L. monocytogens* d) *E. coli*

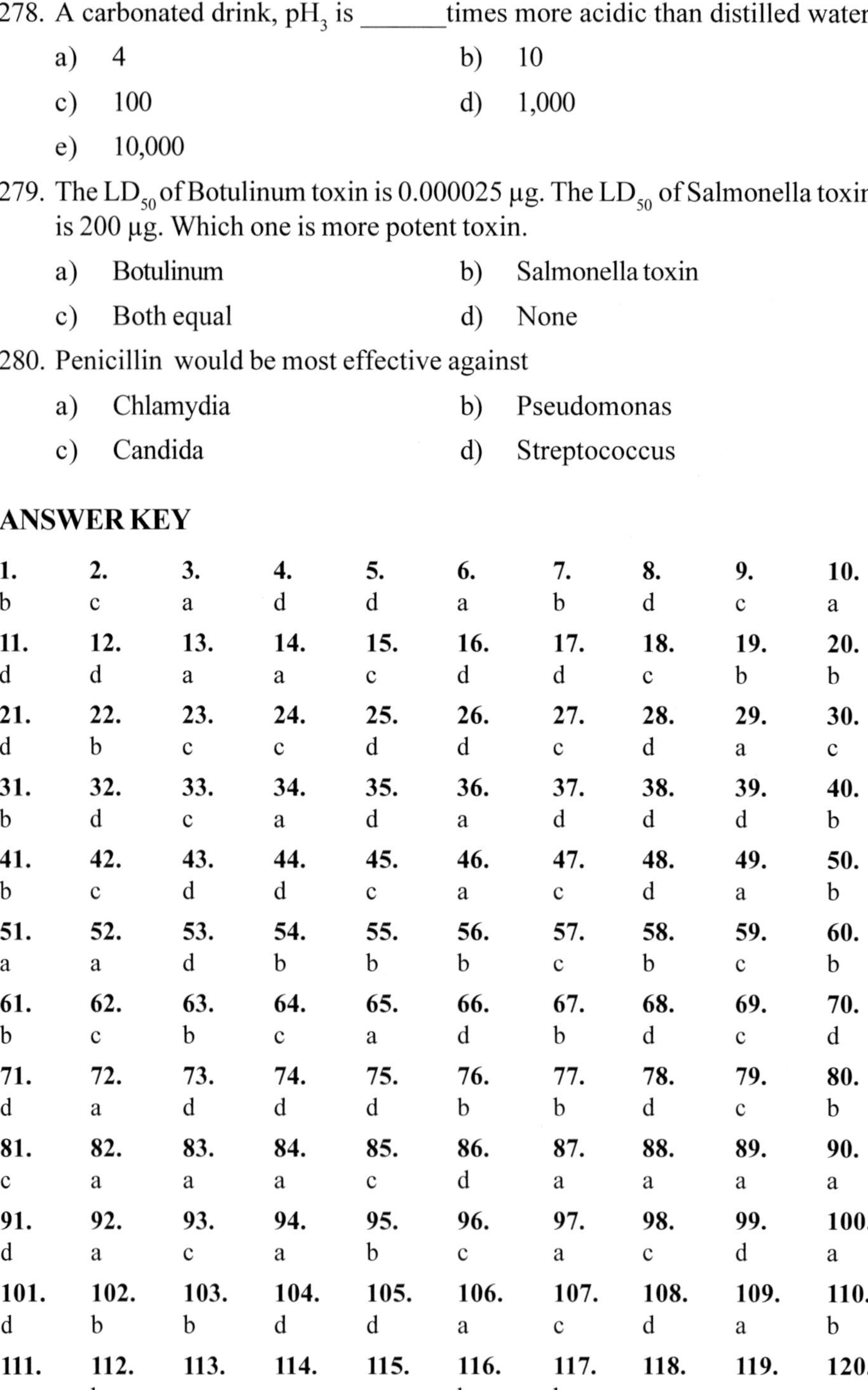

278. A carbonated drink, pH_3 is ______times more acidic than distilled water

  a) 4 b) 10

  c) 100 d) 1,000

  e) 10,000

279. The LD_{50} of Botulinum toxin is 0.000025 µg. The LD_{50} of Salmonella toxin is 200 µg. Which one is more potent toxin.

  a) Botulinum b) Salmonella toxin

  c) Both equal d) None

280. Penicillin would be most effective against

  a) Chlamydia b) Pseudomonas

  c) Candida d) Streptococcus

ANSWER KEY

1.	**2.**	**3.**	**4.**	**5.**	**6.**	**7.**	**8.**	**9.**	**10.**
b	c	a	d	d	a	b	d	c	a
11.	**12.**	**13.**	**14.**	**15.**	**16.**	**17.**	**18.**	**19.**	**20.**
d	d	a	a	c	d	d	c	b	b
21.	**22.**	**23.**	**24.**	**25.**	**26.**	**27.**	**28.**	**29.**	**30.**
d	b	c	c	d	d	c	d	a	c
31.	**32.**	**33.**	**34.**	**35.**	**36.**	**37.**	**38.**	**39.**	**40.**
b	d	c	a	d	a	d	d	d	b
41.	**42.**	**43.**	**44.**	**45.**	**46.**	**47.**	**48.**	**49.**	**50.**
b	c	d	d	c	a	c	d	a	b
51.	**52.**	**53.**	**54.**	**55.**	**56.**	**57.**	**58.**	**59.**	**60.**
a	a	d	b	b	b	c	b	c	b
61.	**62.**	**63.**	**64.**	**65.**	**66.**	**67.**	**68.**	**69.**	**70.**
b	c	b	c	a	d	b	d	c	d
71.	**72.**	**73.**	**74.**	**75.**	**76.**	**77.**	**78.**	**79.**	**80.**
d	a	d	d	d	b	b	d	c	b
81.	**82.**	**83.**	**84.**	**85.**	**86.**	**87.**	**88.**	**89.**	**90.**
c	a	a	a	c	d	a	a	a	a
91.	**92.**	**93.**	**94.**	**95.**	**96.**	**97.**	**98.**	**99.**	**100.**
d	a	c	a	b	c	a	c	d	a
101.	**102.**	**103.**	**104.**	**105.**	**106.**	**107.**	**108.**	**109.**	**110.**
d	b	b	d	d	a	c	d	a	b
111.	**112.**	**113.**	**114.**	**115.**	**116.**	**117.**	**118.**	**119.**	**120.**
c	b	a	a	c	b	b	c	c	a

121.	**122.**	**123.**	**124.**	**125.**	**126.**	**127.**	**128.**	**129.**	**130.**
b	b	d	b	d	d	a	c	b	d
131.	**132.**	**133.**	**134.**	**135.**	**136.**	**137.**	**138.**	**139.**	**140.**
d	b	a	a	a	b	a	a	a	c
141.	**142.**	**143.**	**144.**	**145.**	**146.**	**147.**	**148.**	**149.**	**150.**
a	c	b	c	c	a	d	b	b	a
151.	**152.**	**153.**	**154.**	**155.**	**156.**	**157.**	**158.**	**159.**	**160.**
c	c	b	d	a	c	d	d	a	b
161.	**162.**	**163.**	**164.**	**165.**	**166.**	**167.**	**168.**	**169.**	**170.**
d	a	b	b	d	a	a	d	d	a
171.	**172.**	**173.**	**174.**	**175.**	**176.**	**177.**	**178.**	**179.**	**180.**
d	d	d	c	c	a	d	d	a	c
181.	**182.**	**183.**	**184.**	**185.**	**186.**	**187.**	**188.**	**189.**	**190.**
d	d	d	d	d	b	c	c	b	d
191.	**192.**	**193.**	**194.**	**195.**	**196.**	**197.**	**198.**	**199.**	**200.**
d	a	b	b	c	c	d	a	d	a
201.	**202.**	**203.**	**204.**	**205.**	**206.**	**207.**	**208.**	**209.**	**210.**
d	d	d	a	c	b	b	b	a	b
211.	**212.**	**213.**	**214.**	**215.**	**216.**	**217.**	**218.**	**219.**	**220.**
a	a	c	b	a	d	b	b	d	b
221.	**222.**	**223.**	**224.**	**225.**	**226.**	**227.**	**228.**	**229.**	**230.**
c	d	b	a	a	d	a	c	a	b
231.	**232.**	**233.**	**234.**	**235.**	**236.**	**237.**	**238.**	**239.**	**240.**
d	d	c	c	b	d	c	b	d	d
241.	**242.**	**243.**	**244.**	**245.**	**246.**	**247.**	**248.**	**249.**	**250.**
a	d	a	d	c	a	c	d	a	d
251.	**252.**	**253.**	**254.**	**255.**	**256.**	**257.**	**258.**	**259.**	**260.**
a	a	e	d	c	c	b	b	b	b
261.	**262.**	**263.**	**264.**	**265.**	**266.**	**267.**	**268.**	**269.**	**270.**
c	b	b	c	b	a	a	c	c	c
271.	**272.**	**273.**	**274.**	**275**	**276.**	**277.**	**278.**	**279.**	**280.**
a	c	a	c	e	c	c	e	a	d

6

Food Microbiology-V

1. Nisin is produced by strains of

   a) Streptococcus lactis b) Pseudomonas

   c) E. coli d) Clostridium

2. The inner parts of whole healthy tissues of living plants and animals are

   a) Low in microbial content b) Sterile

   c) High in microbial content d) Both a & b

3. Pseudomonas grows well in food containing

   a) Vitamins b) Organic acids

   c) Antibiotics d) Nitrates

4. Many kinds of molds are

   a) Proteolytic b) Inhibit breakdown of proteins

   c) Lipolytic d) None of these

5. Proteolytic organisms utilize hydrolysis products of

   a) Proteins b) Peptides

   c) Amino acids d) All of the above

6. Osmophilic organisms like yeasts grow best in

   a) Low concentrations of sugar b) High concentrations of sugar

   c) Low concentrations of salt d) High concentrations of salt

7. The aw for pure water is

   a) 1.00 b) 9.99

   c) 0.9823 d) 0.1

8. Organisms that grow over a wide range of pH are

   a) Bacteria b) Yeasts

   c) Thermophillic anaerobes d) Molds

9. Avidin is present in
   a) Meat b) Fish
   c) Eggs d) Fruits
10. Browning of sugar syrups results in the production of
   a) Phenols b) Furfural
   c) Alcohol d) Ketones
11. The fungus known as bread mold is
   a) Mucor b) Penicillium
   c) Rhizopus d) Aspergillus
12. The organism used in the 'amylo' process for the saccharification of starch is
   a) P. digitatum b) A. niger
   c) R. stolonifer d) M. rouxii
13. Trichothecium is commonly called
   a) Pink mold b) Black mold
   c) Red mold d) Green mold
14. Dairy mold is
   a) Trichothecium b) Geotrichum
   c) Thamnidium d) Penicillium
15. The genus Photobacterium causes
   a) Phosphorescence of meat b) Spoilage of cheese
   c) Spoilage of alcoholic beverages d) Stormy fermentation of milk
16. The yellow-orange pigment of which genus causes discoloration on the surface of meat?
   a) Flavobacterium b) Gluconobacter
   c) Escherichia d) Erwinia
17. E. carotovora causes
   a) Ropiness in beer b) Bacterial soft rot
   c) Discoloration of foods high in salt d) Ripening of cheese

18. Aeromonas hydrophila is a
    a) Plant pathogen
    b) Spoilage organism of fruits
    c) Human pathogen
    d) None of the above
19. Red pigmentation on spoilt food is caused by
    a) Pseudomonas
    b) Clostridium
    c) Salmonella
    d) Serratia
20. Pigmented Propionibacter causes
    a) Color defects in breads
    b) Color defects in juices
    c) Color defects in cheese
    d) Color defects in meat
21. Carcasses of animals may be contaminated with
    a) Mycobacterium
    b) Pediococcus
    c) Proteus
    d) Salmonella
22. The number of mciroorganisms in air depends on
    a) Sunshine
    b) Humidity
    c) Amount of suspended dust
    d) All the above
23. The microbial load shed by a human being per minute is
    a) $10^3 - 10^4$
    b) $10^1 - 10^3$
    c) $10^6 - 10^8$
    d) $10^4 - 10^5$
24. During processing, foods can get contaminated because of
    a) Workers
    b) Equipment
    c) Packaging materials
    d) All of them
25. Which of the following is based on the review of air borne contamination by Heldman (1974)?
    a) The microbial populations are different in processing plants.
    b) The population levels are not related to the activity of the workers.
    c) Populations in a plant are not related to the air quality outside the plant.
    d) The microbial populations of different processing plants are similar.
26. Water used for soft drinks are treated with
    a) Sand filters
    b) Chlorination
    c) UV rays
    d) Both b and c

27. The bacterium causing the surface taint of cheese, Pseudomonas putrifaciens,
    - a) Comes primarily from water
    - b) Comes from air
    - c) Comes from sewage
    - d) Comes from soil
28. An important source of heat-resistant spore-forming bacteria is from
    - a) Air
    - b) Sewage
    - c) Soil
    - d) Water
29. Coliform bacteria, anaerobes, Enterococci, other intestinal bacteria and viruses are generally
    - a) Present in soil
    - b) Present in sewage
    - c) Present in water
    - d) Present in animals
30. Mechanical damage to fruits and vegetables by birds open the way
    - a) To microbial spoilage
    - b) To chemical spoilage
    - c) To spread infectious diseases
    - d) None of the above
31. An example of semi-perishable food is
    - a) Potatoes
    - b) Sugar
    - c) Meat
    - d) Eggs
32. Spoilage in foods may be due to
    - a) Insects
    - b) Physical changes
    - c) Growth and activity of microorganisms
    - d) All of the above
33. Criteria for the assurance of fitness in foods is
    - a) Freedom from pollution at any stage in production and handling
    - b) The desired stage of development or maturity
    - c) Both a and b
    - d) None of them
34. Streptococcus lactis causes
    - a) Color change in milk
    - b) No color change in milk
    - c) No spoilage of milk
    - d) Production of antibiotics
35. A dry food like bread is most likely to be spoilt by
    - a) Molds
    - b) Mesophillic bacteria
    - c) Yeasts
    - d) Thermophillic bacteria

36. Freezing prevents
    a) Concentration of solutes b) Microbial growth
    c) Damage to tissues d) Removal of water
37. Carbonates occurring in foods is due to the
    a) Hydrolyzation of organic acids b) Hydroxylation of organic acids
    c) Reduction of organic acids d) Oxidation of organic acids
38. Pectinesterases causes the hydrolysis of
    a) Methyl ester linkage of pectin to yield pectic acid
    b) Pectic acid to pectin
    c) Protopectin to pectic acid
    d) Protopectin to pectin
39. Fats are hydrolysed by
    a) Hydrolases b) Oxidases
    c) Microbial lipase d) Pectinases
40. Desulfotomaculum nigrificans can reduce
    a) Sulfur to sulfide b) Sulfate to sulfide
    c) Sulfide to sulfur d) None of them
41. Asepsis is
    a) Reducing the populations of microorganisms
    b) Identifying methods of destruction
    c) Sterility checks
    d) Keeping out microorganisms
42. Delay in microbial decomposition can be achieved by
    a) Asepsis b) Filtration
    c) Radiation d) All the above
43. Prevention of self-decomposition can be accomplished by
    a) Blanching b) Washing
    c) Chemicals d) Heat
44. Which of the following is not a preservation factor?
    a) To restrict oxygen b) Radiation
    c) Heat d) Mechanical damage

45. The rapid and constant rate of multiplication of an organism occurs during the
    a) Lag phase
    b) Exponential phase
    c) Stationary phase
    d) Survival phase
46. Microbial decomposition of foods can be prevented by
    a) Recontamination
    b) Killing microbes
    c) Scalding
    d) None of them
47. Restriction of access of microorganisms to a product cannot be achieved by using
    a) Heat
    b) HCl
    c) H_2O_2
    d) H_2O
48. The widely used application of asepsis is
    a) Clean handling
    b) Decontamination
    c) Packaging
    d) Radiation
49. Heat is used to
    a) Inactivate microbes
    b) Inhibit growth of microbes
    c) Kill microbes
    d) Restrict the growth of microbes
50. 'Bacteria-proof' filter is made of
    a) Sintered glass
    b) Diatomaceous earth
    c) Unglazed porcelain
    d) All the above
51. Berries and Sauerkraut are classified as
    a) High-acid foods
    b) Acid foods
    c) Medium-acid foods
    d) Low-acid foods
52. Moist heat at 60°C for 5 to 10 minutes kills most
    a) Thermophillic bacteria
    b) Yeasts
    c) Sclerotia
    d) Molds
53. TDT is
    a) Thermal death transfer
    b) Transferred death time
    c) Thermal death time
    d) Thermal delayed time
54. Heat penetration in processing can be hastened by
    a) Using glass containers
    b) Rotation and agitation
    c) Layering of foods
    d) Using larger cans

55. The 'Father of canning is'
    a) Nicholas Appert b) Spallanzani
    c) Lopez d) Peter Durand
56. Enamel coating in cans containing zinc oxide cannot be used to store
    a) Berries b) Beets
    c) Meat d) Corn
57. Propellant gas used for pressurized packaged foods include
    a) Oxygen b) Carbon dioxide
    c) Hydrogen sulfide d) Sulfur gas
58. Heating of cans can be done
    a) In a boiling water bath b) Using a steamer
    c) In oven heat d) All the above
59. Meat, fish, etc., are grouped under
    a) Low-acid foods b) Medium-acid foods
    c) Acid foods d) High-acid foods
60. An exposure at 60°C for 10 to 15 minutes kills
    a) Ascospores b) Oospores
    c) Basidiospores d) None of them
61. The temperature in cellar storage is lower than
    a) 15°C b) 5°C
    c) 1°C d) 25°C
62. Freezer burn is produced when
    a) Thawing is not done b) Ice crystals melt
    c) Foods are desiccated d) Ice crystals evaporate from the surface
63. Lethal effects are caused due to
    a) Freezing at 4°C b) Flocculation of essential proteins
    c) Freezing using liquid nitrogen d) Both c and d
64. Storage death refers to
    a) Death and gradual reduction in microorganisms
    b) Death of cells
    c) Death of tissues
    d) Death of cells and reduction in nutritional levels

65. Cold shock results in
    a) Alterations of lipids in the membrane
    b) Increased permeability of membranes
    c) Syneresis
    d) Increased microbial count
66. The process of freezing foods using liquid nitrogen is called
    a) Lyophillization b) Dehydro-freezing
    c) Cryoinjury d) Sharp freezing
67. The rapid slowing of chemical and enzymatic reactions in foods is achieved by
    a) Quick freezing b) Sharp freezing
    c) Dehydro-freezing d) None of them
68. The unfrozen, concentrated solutions of sugars and salt produces a viscous material called
    a) Osmotic solution b) Liquid water
    c) Metacryotic liquid d) None of them
69. Mushiness of fruits on thawing results in physical damage during
    a) Freezing b) Air-blast freezing
    c) Immersion in water d) Storage in ice
70. Chilling storage enables the preservation of products for
    a) A longer period of time b) A limited period of time
    c) A period of two months d) None of them
71. Which of the following is not a factor in the control of drying?
    a) The temperature employed b) The relative humidity of the air
    c) The velocity of the air d) The concentration of heat
72. Thermoduric streptococci are predominant in
    a) Dried milk b) Fruits and nuts
    c) Spoiled eggs d) Vegetables
73. 99% of bacteria are removed by
    a) Blanching b) Sweating
    c) Smoking d) Sun drying

74. Sulfuring helps
    a) Conserve vitamin C
    b) Maintain an attractive color
    c) Repel insects
    d) All of the above
75. 'Sweating' is storage for
    a) Re-addition of moisture to a desired level
    b) Reduces moisture
    c) Increases moisture
    d) None of the above
76. Foam mat drying is used for
    a) Liquid food
    b) Powdered food
    c) Dehydrated food
    d) Dried fruits
77. The simplest dryer is the
    a) Kiln
    b) Oven
    c) Sun
    d) Microwave
78. Sodium carbonate is used as an alkali dip before sun drying for
    a) Cereals
    b) Nuts
    c) Vegetables
    d) Grapes
79. Intermediate-moisture foods are those
    a) That contain 20–40% moisture
    b) That have non-refrigerated shelf life
    c) That can be eaten without preparation
    d) All the above are true
80. The chief bacteria on dried vegetables are
    a) Enterobacter
    b) Bacillus
    c) Clostridium
    d) all of them
81. Sorbic acid is known
    a) To inhibit yeasts and molds
    b) To destroy insects
    c) To remove bacteria
    d) To eliminate disease-causing agents
82. Vinegar is used in
    a) Mayonnaise
    b) Pickled sausages
    c) Pickles
    d) All the above

83. Wood smoke contains
    a) Compounds deleterious to molds
    b) Active substances that reduce yeasts
    c) Volatile compounds that are bacteriostatic
    d) Flavoring compounds that have no effects on foods
84. Washing foods or equipment is achieved by
    a) Adding acids b) Adding halogens
    c) Adding phosphates d) Adding salt water
85. Butyl thiocyanate is an active principle in
    a) Horse radish b) Cabbage
    c) Turnips d) Garlic
86. A mold inhibitor used as a spray to kill airborne microbes is
    a) Caprylic acid b) Sorbic acid
    c) Propylene glycol d) Vanillic acid
87. Fumes of burning sulphur is used to treat
    a) Dark-colored vegetables b) Light-colored fruits
    c) Meat d) Eggs
88. Propionic acid is found naturally in
    a) Butter b) Milk
    c) Malt extract d) Swiss cheese
89. The organic acid used in drinks, jams, jellies, and syrups is
    a) Oxalic acid b) Tartaric acid
    c) Citric acid d) Acetic acid
90. Excess addition of which of the following causes high osmotic pressure with plasmolysis of cells?
    a) Salt b) Sugar
    c) Organic acid d) Both a and b
91. Purines and pyrimidines absorb radiations at wavelengths of
    a) 260 nm b) 200 nm
    c) 250 nm d) 270 nm

92. The effectiveness of UV rays are influenced by
    a) Time b) Intensity
    c) Penetration d) All of the above

93. UV irradiation in the food industry is used for the treatment of
    a) Water for beverages b) Knives for slicing
    c) Only a d) Both a and b

94. The drawback of using X-rays in food preservation is
    a) Their low efficiency b) High cost of production
    c) They are toxic d) Both a and b

95. The rays that are directional and less penetrating are
    a) Gamma rays b) Cathode rays
    c) X-rays d) Beta rays

96. Side reactions due to radiations results in
    a) Undesirable color b) Change in physical properties
    c) Undesirable tastes d) All of the above

97. Low-level irradiation can be used on fresh fruits to
    a) Kill insects b) Kill molds
    c) To inhibit the growth of bacteria d) None of them

98. The general nutrition of an irradiated food will be
    a) As good as that of a food processed by any other means
    b) Not as good as that of a food processed by any other means
    c) Food quality deteriorates
    d) The food attracts spoilage organisms

99. In which food does the destruction of glutathione occur?
    a) Fruits b) Vegetables
    c) Meat d) Milk

100. Spores of C. botulinum have been found to be resistant to
    a) Beta rays b) Gamma rays
    c) Cathode rays d) X-rays

101. The process of milling which reduces the number of microorganisms is
   a) Bleaching b) Washing
   c) Sifting d) Dry cleaning

102. Actinomycetes are the normal flora of
   a) Flour b) Cereals and snacks
   c) Frozen dough d) Raw cereal grains

103. Boston brown bread is
   a) Wrapped b) Canned
   c) Packed d) Exposed

104. The percentage of ammonia used to reduce mold growth in high-moisture corn is
   a) 10% b) 1%
   c) 2% d) 5%

105. The organic acid used to combat rope in bakery products is
   a) Acetic acid b) Citric acid
   c) Tartaric acid d) Fulvic acid

106. The major factors involved in the spoilage of stored grains by molds include
   a) Microbial content b) Moisture levels above 12%
   c) Physical damage d) All of the above

107. Bleaching of white wheat flour is done by using oxidizing agents like
   a) Benzoyl peroxide b) Hydrogen peroxide
   c) Sulfur dioxide d) Chlorobenzene

108 In bread Bacillus subtilis causes
   a) Decay b) Rotting
   c) Ropiness d) Pigmentation

109. Swelling of moist macaroni is caused due to the presence of
   a) Enterobacter cloacae b) Staphylococcus aureus
   c) Lactobacillus cremoris d) Escherichia coli

110. Endomycopsis causes
   a) Chalky bread b) Red bread
   c) Blue bread d) None of the above

111. The source of yeast in honey comes from the
   a) Air
   b) Nectar of flowers
   c) Spoilage of fruits
   d) None of them
112. Two main groups of bacteria present during maturation of nectar to honey are
   a) Gluconobacter and lactobacillus
   b) Pseudomonas and Alcaligenes
   c) Flavobacterium and Bacillus
   d) Leuconostoc and Clostridium
113. Osmophilic yeasts that survive the heat process, cause spoilage of
   a) Liquid sugars
   b) Honey
   c) Candies
   d) Canned molasses
114. Ropy or stringy sap is caused by
   a) Yeasts
   b) Enterobacter aerogenes
   c) Pseudomonas striata
   d) Flavobacterium spp.
115. In maple sap Pseudomonas fluorescences causes
   a) Greenish sap
   b) Yellow sap
   c) Red sap
   d) White sap
116. Commercial honey is pasteurized at
   a) 71–77°C for a few minutes
   b) 71–77°C for 1 hour
   c) 71–77°C for a few days
   d) 71–77°C for 10 hours
117. Fungal growth is inhibited in cane or sugar beet by
   a) 5% CO_2 and 6% O_2
   b) 6% CO and 5% O_2
   c) 6% CO_2 and 5% O_2
   d) 5% CO and 6% O_2
118. Which of the following are easily subjected to deterioration?
   a) Raw cane juice
   b) Sucrose
   c) Honey
   d) Molasses
119. Which of the following statements are not true for sucrose?
   a) The purer the product, the poorer it becomes as a culture medium
   b) The more concentrated it gets, the fewer kinds of organisms can grow in it
   c) The purer the product, the better it becomes as a culture medium
   d) None of the above

120. Chocolate candies have been incriminated in cases of
    a) Salmonellosis
    b) Botulism
    c) Staphylococcal food intoxication
    d) Perfringens poisoning
121. Hydrocooling refers to
    a) Use of cold water spray
    b) Spraying of liquid nitrogen
    c) Ice crystal formation
    d) None of the above
122. Vegetables can be dried by
    a) Sun drying
    b) Explosive puffing
    c) Dehydration
    d) Pressing
123. Fruits are enclosed in wrappers treated with
    a) Borax
    b) Sulfite
    c) Iodine
    d) All of the above
124. Anthracnose is caused by
    a) Alternaria
    b) Erwinia amylovora
    c) Colletotrichum lindemuthianum
    d) Penicillium digitatum
125. Sprouting of potatoes can be delayed by
    a) Dehydration
    b) Drying
    c) Exposure to high temperatures
    d) Irradiation
126. Sulphite residues in foods may be associated with
    a) Asthmatic attacks
    b) Heart attacks
    c) Kidney failure
    d) Brain damage
127. Clear fruit juices are sterilized by
    a) Filtration
    b) Heating
    c) Canning
    d) Nne of the above
128. Mechanical protection to fruits can be enhanced by the use of
    a) Waxed wraps
    b) Paraffin oil
    c) Mineral oil
    d) All of them

129. To double the storage time of loosely packed small fresh fruits, these fruits are exposed to
   a) Ozone b) Carbon dioxide
   c) Oxygen d) Nitrogen

130. Gray mold rot caused by Botrytis is favored by
   a) High humidity and warm temperature
   b) High humidity and cold temperatures
   c) Low humidity and warm temperatures
   d) None of the above

131. Surface slime in meat can be caused by
   a) Moraxella b) Acinetobacter
   c) Micrococcus d) All of the above

132. Puffers or gassy hams is seen when
   a) Inexpert curing is done b) Improper refrigeration
   c) Bacterial contamination d) Reduced heat

133. Dried beef or beef hams become red due to the presence of
   a) Vibrio vulnificus b) Serratia marcescens
   c) Halobacterium salinarium d) Streptococcus lactis

134. Bone taint refers to
   a) Putrifaction next to the bone b) Decay of the tendons
   c) Loss of calcium in the bone d) None of the above

135. Thamnidium chaetocladioides in meat produces
   a) Ropiness b) Whiskers
   c) Discoloration d) Phosphorescence

136. Sodium nitrate a bacteriostatic in acid solution is effective against
   a) Aerobes b) Thermophiles
   c) Acidophiles d) Anaerobes

137. Nitrates play a role in the color of
   a) Fruits b) Vegetables
   c) Meat d) Sweet meat

138. The factors influencing the invasion of microbes in meat tissues are
    a) The load in the gut of the animal
    b) The method of killing and bleeding
    c) The physiological condition of the animal after slaughter
    d) All of the above

139. Green patches in meat are caused by
    a) P. expansum b) P. putida
    c) P. syncyanea d) P vulgaris

140. Packaging films with poor gas penetration encourage the growth of
    a) Bacillus b) Lactic acid bacteria
    c) Clostridium d) Mucor

141. Salt fish are spoiled by
    a) Acidophilic bacteria b) Halophilic bacteria
    c) Thermophilic bacteria d) Psychrophilic bacteria

142. Pseudomonas fluorescens causes spoilage of fish indicated by
    a) Yellow to greenish yellow color
    b) Greenish blue color
    c) Reddish tinge
    d) No color change

143. The kind and rate of spoilage of fish vary with
    a) The kind of fish b) Temperature
    c) The condition of the fish when caught d) All of the above

144. What is germicidal ice?
    a) Ice with chemical preservatives
    b) Ice which is crushed
    c) Ice with antiseptic solution
    d) None of the above

145. The pH of the flesh after the death of fishes is related to the amount of
    a) Oxygen available at death b) Protein content of the flesh
    c) Bacteria in the flesh d) Glycogen available at death

146. Fish can be preserved by

a) Freezing b) Canning

c) Drying d) All of the above

147. Spoilage of smoked fish can be delayed using

a) Acetic acid b) Sorbic acid

c) Benzoic acid d) Boric acid

148. Antioxidants to preserve fishes are used as

a) Glazes b) Gases

c) Dips d) All of the above

149. Deterioration of fatty fish produces appreciable amounts of 'stale fishy', which is

a) Trimethylamine b) Chloramines

c) Ammonia d) Unsaturated fatty acids

150. Chocolate-brown discoloration in fish is caused by

a) Serratia b) Bacillus

c) Proteus d) Asporogenous yeast

151. Defects in fresh egg include

a) Bloom b) Meat spots

c) Cracks d) All of them

152. Green rots in eggs is caused by

a) P. fluorescens b) P. graveolens

c) Achromobacterperolens d) Paracolobactrum

153. Proteus causes spoilage of eggs by producing

a) Red rots b) Colorless rot

c) Black rots d) Pink rot

154. Hay odor in fishes is caused by

a) Sporotrichum b) Enterobacter cloacae

c) P. fluorescens d) Cladosporium

155. The final stage of spoilage of eggs by molds is

a) Black rot b) Pin-spot molding

c) Fungal rotting d) Superficial fungal spoilage

156. Untreated eggs lose moisture during storage and hence

a) Lose weight b) Become heavy

c) Get discolored d) None of the above

157. Dry packing of eggs are done by using

a) Salt and sand b) Lime and sawdust

c) Oiling and waxing d) Both a and b

158. Sodium pentachloro phenate inhibits

a) Molds b) Yeasts

c) Bacteria d) Virus

159. Thermostabilization is a method of dipping eggs into hot water to

a) Wash the eggs b) Increase the water content

c) Reduce evaporation of moisture d) Reduce contamination of eggs

160. Eggs are selected for storage by

a) Waxing b) Oiling

c) Candling d) None of them

161. Increase in concentrations of carbon dioxide in the atmosphere of stored chicken inhibits the growth of

a) Psychrotrophs b) Mesotrophs

c) Thermophiles d) Alkalophiles

162. To reduce the number of organisms in a chill tank

a) Calcium is added b) Carbon dioxide is added

c) Mercury is added d) Chlorine is added

163. Most poultry are preserved by

a) Chilling b) Freezing

c) Heat treatment d) Both a and b

164. An 'outside cut' is

a) If the trachea is cut from inside

b) If the trachea is partially cut

c) If the trachea is left intact

d) None of the above

165. Soaking cut-up poultry in solutions of organic acids is to

a) Lengthen shelf life
b) Shorten shelf life
c) Increase the bacterial count
d) Reduce the bacterial count

166. The pigment pyoverdine in meat is produced by

a) Serratia
b) Pseudomonas
c) Bacillus
d) Vibrio

167. Microaerophillic bacteria are seen in

a) Chicken wrapped in oxygen-impermeable films
b) Vacuum-packed chicken
c) White meat
d) Dark meat

168. Vacuum-packed chicken contains spoilage organisms like

a) Enterobacter
b) Pseudomonas
c) Clostridium
d) Bacillus

169. Pseudomonas, the chief spoilage organism of cut-up poultry, develops a slime accompanied by

a) Taint
b) Acidic smell
c) Sour taste
d) All the above

170. Chicken carcasses are irradiated to effectively destroy

a) Salmonella
b) Enterococci
c) Clostridium
d) Pseudomonas

171. 'Rabbito' is also called

a) Flavor enhancer
b) Sweaty feet
c) Fishiness
d) Surface taint

172. Black smudge in salted butter is caused by

a) Phoma
b) Pseudomonas nigrificans
c) P. fluorescens
d) Penicillium

173. Aeromonas hydrophila causes fishiness in

a) Milk
b) Cheese
c) Butter
d) Cream

174. Pseudomonas syncyanea causes

a) Yellow milk b) Red milk

c) Brown milk d) Blue milk

175. Inhibitory substances present in milk are

a) Lactoperoxidase and agglutunins

b) Renin and casein

c) Lactose and fats

d) None of them

176. Stormy fermentation of milk results in

a) Foam at the top if the milk is liquid

b) Gas bubbles caught in the curd

c) Floating curd containing gas bubbles

d) All of the above

177. Sweet curdling occurs at an early stage of

a) Proteolysis b) Souring

c) Lipolysis d) Curdling

178. Evaporated milk is made by removing

a) 50% of water from whole milk

b) 40% of water from whole milk

c) 60% of water from whole milk

d) 80% of water from whole milk

179. Rapid heating of cream, accompanied by injecting steam is known as

a) Pasteurization b) Vacreation

c) Sterilization d) Cooking

180. The centrifugal procedure used for removing bacteria from milk is

a) Centrifugation b) Bactofugation

c) Separation d) None of them

181. Soft swell has

a) Both ends of the can bulged b) Distorted ends of the can

c) Flat ends of the can d) Dented seams of the can

182. Hydrogen swells is favored by
    a) Increased acidities in foods
    b) Increasing temperature of storage
    c) A poor exhaust
    d) All of the above

183. Bacillus coagulans causes
    a) Flat sour spoilage   b) TA spoilage
    c) Sulfide spoilage   d) None of them

184. TA spoilage is caused by
    a) *C. perfringens*   b) *C. botulinum*
    c) *C. nigrificans*   d) *C. thermosaccharolyticum*

185. Molds found growing in jellies and candied fruits can grow in sugar concentrations up to
    a) 67.5%   b) 60%
    c) 67%   d) 65%

186. Black beets caused by Bacillus beta nigrificans occurs in the presence of
    a) Low salt content   b) High salt content
    c) High content of soluble iron   d) Low content of soluble iron

187. 'Sulphide stinker' is caused by
    a) *C. thermosaccharolyticum*   b) *B. coagulans*
    c) *C. butyricum*   d) Desulfotomaculum nigrificans

188. Canned hams often show the presence of
    a) *S. faecalis*   b) *Lactobacillus*
    c) *Leuconostoc*   d) *Bacillus*

189. Byssochlamys fulva in canned fruits produces
    a) Sclerotia   b) Ascospores
    c) Chlamydospores   d) Oospores

190. Putrid swells of meat is caused by
    a) *C. sporogenes*   b) *C. pasteurianum*
    c) *C. putrefaciens*   d) *C. butyricum*

191. Yellow pigmented Micrococci and Bacilli cause stamping ink which is

a) Discoloration of bones
b) Discoloration of meat fat
c) Discoloration of hooves
d) Discoloration of hide

192. Spoilage of soft drinks is evidenced by

a) Rancidity and sourness
b) Offensive odor and taste
c) Cloudiness and ropiness
d) Discoloration and clarity

193. Carbonation is inhibitory and

a) Germicidal
b) Antiseptic
c) Bactericidal
d) None of these

194. Musty odor and taste in root beer is caused by

a) Clostridium
b) Achromobacter
c) Bacillus
d) Aspergillus

195. Solar salt is obtained from the

a) Evaporation of surface salt water
b) Sun drying of salt
c) Filtration of water
d) Heating in furnaces

196. Halobacterium salinarium is present in

a) Wet salt
b) Purified salt
c) Welled salt
d) Solar salt

197. Undesirable flavor in fats or oils is a result of

a) Souring
b) Cloudiness
c) Rancidity
d) Contamination

198. As a result of oxidation, butter fat and meat fat become

a) Tallowy
b) Discolored
c) Supple
d) Dehydrated

199. The content of microorganisms in spices can be reduced with a treatment of

a) Alcohol
b) Glutaraldehyde
c) Propylene oxide
d) Benzoic acid

200. Growth of Clostridium botulinum is seen in
   a) Imitation cheese
   b) Fresh mushroom
   c) Both a and b
   d) None

201. The microorganisms added to Swiss cheese to improve flavor and assist eye formation is
   a) *Lactobacillus cremoris*
   b) *Lactobacillus bulgaricus*
   c) *Streptococcus thermophilus*
   d) *Propionibacterium freudenreichii*

202. Moistened sterile crackers support the growth of
   a) *P. camemberti*
   b) *A. niger*
   c) *R. stolonifer*
   d) *P. chrysogenum*

203. Distiller's yeast is a high-alcohol yielding strain of
   a) Candida
   b) Torula
   c) S. cerivisiae
   d) Torulopsis

204. Soil stocks are preserved by
   a) Freeze drying
   b) Glycerol stocks
   c) Drying
   d) Heat fixing

205. Dried yeast contains
   a) 8% moisture
   b) 5% moisture
   c) 10% moisture
   d) 0% moisture

206. Baker's yeast can also be prepared from
   a) Grain mashes
   b) Waste sulfite liquor
   c) Wood hydrolysate
   d) A ll the above

207. Impure mixed cultures are required for the production of
   a) Citric acid
   b) Lactic acid
   c) Cinegar
   d) Alcohol

208. Bacterial cultures have been preserved for months at room temperature on agar slants which contain
   a) 5% NaCl
   b) 1% NaCl
   c) 2% NaCl
   d) 4% NaCl

209. The activity of a culture is judged by its
   a) Rate of growth
   b) Production of products
   c) Both a and b
   d) None of the above

210. Deterioration of cultures may result from

a) Improper handling

b) Cultivation

c) Frequent transfer over long periods in an inadequate culture medium

d) All the above

211. 'Wort' is a

a) Dense solution b) Clear solution

c) Solution with suspended particles d) None of them

212. Boiling of the wort with hops is to

a) Inactivate enzymes b) Sterilize it

c) To caramelize sugar slightly d) All of the above

213. 'Sarcina sickness' of beer is caused by

a) Saccharomyces cerevisiae b) Pediococcus cerevisiae

c) Lactobacillus pastorianus d) Hansenulaanomala

214. Silky turbidity in beer is caused by

a) Zymomonas anaerobium b) Acetobacter pasteurianus

c) Gluconobacteroxydans d) Lactobacillus diastaticus

215. Sake a yellow rice beer has an alcohol content of

a) 5-10% b) 14-17%

c) 20-25% d) 2-5%

216. An undesirable process called aceti- ficaion involving the oxidation of alcohol is caused by

a) Streptococcus mucilaginosus b) Zymomonasanaeobium

c) Gluconobacteroxydans d) Lactobacillus diastaticus

217. Fermentation of glycerol in wine results in

a) Pousse b) Amertume

c) Mousy flavor d) T ourne

218. 4% commercially available acetic acid is

a) Brine b) Vinegar

c) Tartar d) Salt

219. 'Fish Eye' spoilage of ripe olives is caused by

a) Bacillus subtilis b) Bacillus pumilus

c) Bacillus polymyxa d) only a and b

220. Slimy or ropy kraut is caused by

a) Lactobacillus plantarum b) Erwinia herbicola

c) Leuconostoc mesentroides d) Lactobacillus brevis

221. SCP can be obtained from

a) Yeasts b) Fungi

c) Algae d) All of the above

222. Dextran can be obtained from

a) Molasses b) Sugar beet

c) Parafin d) Ethanol

223. A. niger is the principal mold used in the production of

a) Oxalic acid b) Acetic acid

c) Amino acids d) Citric acid

224. The most important organism in the production of amylase is

a) Bacillus subtilis b) Candida pulcherima

c) Geotrichumcandidum d) Saccharomyces cerevisiae

225. Invertase is used in the manufacture of

a) Confectionery b) Fruit juices

c) Jams d) Jellies

226. Pectinases are used in food industries to

a) Enhance flavor b) Clarify fruit juices

c) Enhance color of juices d) Retain freshness of juices

227. Addition of which acid makes milk more digestible to infants?

a) Citric acid b) Gluconic acid

c) Amino acid d) Lactic acid

228. The substrate used in the production of SCP is

a) Molasses b) Acid hydrolysate of wood

c) Fruit wastes d) All the above

229. The limitations on the use of bacteria as SCP is
   a) Poor public acceptance b) Small size
   c) High content of nucleic acids d) All the above
230. S. maxima is grown commercially in
   a) Lake Texcoco b) Lake Victoria
   c) Lake Missisipi d) None of the above
231. L. monocytogenes can be isolated from
   a) Milk b) Sewage
   c) Soil d) Both a and b
232. The diarrheal syndrome and the emetic syndrome are characteristic of
   a) Staphylococcal food poisoning b) Salmonellosis
   c) Perfringens poisoning d) Bacillus cereus food poisoning
233. Sea foods and sea water may contain
   a) Vibrio vulnificus b) Streptococcus faecalis
   c) Aeromonas hydrophilia d) Vibrio parahaemolyticus
234. A characteristic cysteine loop in the structure of the enterotoxin is seen in the toxin produced by
   a) Salmonella b) Staphylococcus
   c) Yersinia d) Bacillus
235. Kauffman-White scheme identifies isolates of
   a) Salmonella b) Clostridium
   c) Staphylococcus d) None of the above
236. Food poisoning caused by a gram negative, curved motile rod identified in Japan is
   a) V. parahaemolyticus b) Salmonella typhimurium
   c) Escherichia coli d) Shigella sonnei
237. C. jejuni has been isolated from
   a) Chicken carcasses b) Turkeys
   c) Pork sausages d) All of the above
238. Aeromonas hydrophilia may be pathogenic to
   a) Frogs b) Ducks
   c) Rabbits d) None of the above

239. Traveler's diarrhea is caused due to the toxins produced by

a) Clostridium b) Bacillus

c) Escherichia d) Yersinia

240. Scarlet fever and septic sore throat are diseases caused by

a) Bacillus cereus b) Streptococcus pyogenes

c) Arizona hinshawii d) Shigella boydii

241. The syndrome resulting from the ingestion of toxin in a mold contaminated food is

a) Aspergillosis b) Enterotoxicosis

c) Mycotoxicosis d) Neurotoxicosis

242. Patulin is toxic to

a) Mice b) Fishes

c) Cattle d) Human beings

243. Alimentary toxic aleukia isolated from grain is produced by

a) Alternaria b) Cladosporium

c) Fusarium d) All of the above

244. Mold nephrosis, a disease in pigs has been associated with

a) Aflatoxin b) Patulin

c) Citrinin d) Ochratoxin

245. Scombroid fish poisoning has been associated with

a) Proteus morganii b) Klebsiella pneumonia

c) Lyngbyamajuscula d) Both a and b

246. Lyngbya majuscule causes

a) Ciguatera poisoning b) Scombroid poisoning

c) Shellfish poisoning d) None of the above

247. A toxic substance detected in blue cheese is

a) Luteoskyrin b) Roquefortine

c) Patulin d) Ochratoxin

248. Penicillium islandicum produces

a) Citrinin b) Sterigmatocystin

c) Luteoskyrin d) Patulin

249. Aflatoxins are produced by

a) A. flavus b) A. parasiticus
c) A. niger d) Both a and b

250. A chlorinated isocoumarin derivative is

a) Ochratoxin b) Aflatoxin
c) Patulin d) Citirinin

251. V. parahaemolyticus has an incubation period of

a) 15-24 hours b) 30 min-8 hours
c) 12-50 hours d) 6-48 hours

252. Soft drinks are tested for chemicals when they are contained in a

a) Glass bottle b) Tetrapack
c) Metal container d) None of them

253. The toxic substance of the 'Chinese restaurant' syndrome is

a) Acetic acid b) MSG
c) Vinegar d) Chilli sauce

254. Corynebacterium is detected by the analysis of

a) Throat swab b) Urine sample
c) Blood sample d) Skin lesions

255. Samples of suspected perishable foods must be kept cold

a) During collection of the sample
b) During analysis
c) During transit to laboratory
d) There is no need to keep the samples cold

256. Cultures from nose, throat or skin lesions are used to test for

a) Salmomnella b) Bacillus
c) Clostridium d) Staphylococcus

257. Mucoid diarrhea is the predominant symptom in the detection of

a) Giardia lamblia b) Enteric viruses
c) Entamoeba histolytica d) Taenia solium

258. Neurological symptoms caused by puffer-fish poisoning is because of

a) Shellfish toxin b) Tetraodon toxin
c) Both a and b d) None of them

259. The predominant symptoms in the lower gastro-intestinal tract infection is
   a) Fever b) Abdominal cramps and diarrhea
   c) Chills d) Malaise
260. Cyanosis occurs due to
   a) Phosphate poisoning b) Loss of iron
   c) Nitrite poisoning d) None of the above
261. A system to control the safety of a manufactured product is determined by
   a) HACCP b) ISO 9000
   c) CCP d) FAO
262. HACCP is designed to
   a) Prevent problems before they occur
   b) Correct deviations as soon as detected
   c) Both a and b
   d) None of the above
263. National academy of sciences is an authority on
   a) Food additives b) Food safety
   c) Food preservation d) Food processing
264. The Codex guidelines for HACCP has
   a) Five principles b) Six principles
   c) Seven principles d) None of the above
265. CIP refers to
   a) Clean in place systems b) Cleanliness in place systems
   c) Cean in places systems d) None of them
266. During transportation, perishable foods should be kept at
   a) 3.3–4.4°C b) 66°C
   c) 0°C d) Both a and b
267. An example of a wetting agent is
   a) Polyether alcohol b) Chlorine
   c) Trisodium phosphate d) Citric acid
268. CCP aims at preventing or reducing
   a) Physical hazards b) Chemical hazards
   c) Biological hazards d) All the above

269. Validation ensures that the industry
    a) Complies with the required plan
    b) Has random sampling
    c) Prevents deviations
    d) Lists the significant hazards
270. The simplest record-keeping system to ensure effectiveness is by
    a) Establishing procedures for verification
    b) Establishing documentation
    c) Establishing monitoring systems
    d) Establishing critical control points

ANSWER KEY

1.	**2.**	**3.**	**4.**	**5.**	**6.**	**7.**	**8.**	**9.**	**10.**
a	d	d	a	d	a	a	d	c	b
11.	**12.**	**13.**	**14.**	**15.**	**16.**	**17.**	**18.**	**19.**	**20.**
c	d	a	b	a	a	b	c	d	c
21.	**22.**	**23.**	**24.**	**25.**	**26.**	**27.**	**28.**	**29.**	**30.**
d	d	a	d	d	d	a	c	b	a
31.	**32.**	**33.**	**34.**	**35.**	**36.**	**37.**	**38.**	**39.**	**40.**
a	d	c	b	a	b	d	a	c	b
41.	**42.**	**43.**	**44.**	**45.**	**46.**	**47.**	**48.**	**49.**	**50.**
d	d	a	d	b	b	d	c	a	d
51.	**52.**	**53.**	**54.**	**55.**	**56.**	**57.**	**58.**	**59.**	**60.**
a	d	c	b	a	c	b	d	a	a
61.	**62.**	**63.**	**64.**	**65.**	**66.**	**67.**	**68.**	**69.**	**70.**
a	d	d	a	a	b	a	c	a	b
71.	**72.**	**73.**	**74.**	**75.**	**76.**	**77.**	**78.**	**79.**	**80.**
d	a	a	d	a	a	a	d	d	d
81.	**82.**	**83.**	**84.**	**85.**	**86.**	**87.**	**88.**	**89.**	**90.**
a	d	c	b	a	c	b	d	c	d
91.	**92.**	**93.**	**94.**	**95.**	**96.**	**97.**	**98.**	**99.**	**100.**
a	d	d	d	b	d	a	a	c	b
101.	**102.**	**103.**	**104.**	**105.**	**106.**	**107.**	**108.**	**109.**	**110.**
a	d	b	c	a	d	a	c	a	a
111.	**112.**	**113.**	**114.**	**115.**	**116.**	**117.**	**118.**	**119.**	**120.**
b	a	d	b	a	a	c	a	c	a

121.	**122.**	**123.**	**124.**	**125.**	**126.**	**127.**	**128.**	**129.**	**130.**
a	b	d	c	d	a	a	d	a	a
131.	**132.**	**133.**	**134.**	**135.**	**136.**	**137.**	**138.**	**139.**	**140.**
d	a	c	a	b	d	a	d	a	b
141.	**142.**	**143.**	**144.**	**145.**	**146.**	**147.**	**148.**	**149.**	**150.**
b	a	d	a	d	d	b	d	a	d
151.	**152.**	**153.**	**154.**	**155.**	**156.**	**157.**	**158.**	**159.**	**160.**
d	a	c	a	c	a	d	a	c	c
161.	**162.**	**163.**	**164.**	**165.**	**166.**	**167.**	**168.**	**169.**	**170.**
a	d	d	c	a	b	a	a	d	a
171.	**172.**	**173.**	**174.**	**175.**	**176.**	**177.**	**178.**	**179.**	**180.**
d	b	c	d	a	d	a	c	b	b
181.	**182.**	**183.**	**184.**	**185.**	**186.**	**187.**	**188.**	**189.**	**190.**
a	d	a	d	a	c	d	a	b	a
191.	**192.**	**193.**	**194.**	**195.**	**196.**	**197.**	**198.**	**199.**	**200.**
b	c	a	b	a	d	c	a	c	c
201.	**202.**	**203.**	**204.**	**205.**	**206.**	**207.**	**208.**	**209.**	**210.**
d	a	c	a	a	d	c	b	c	d
211.	**212.**	**213.**	**214.**	**215.**	**216.**	**217.**	**218.**	**219.**	**220.**
b	d	b	a	b	c	b	b	d	a
221.	**222.**	**223.**	**224.**	**225.**	**226.**	**227.**	**228.**	**229.**	**230.**
d	a	d	a	a	b	d	d	d	a
231.	**232.**	**233.**	**234.**	**235.**	**236.**	**237.**	**238.**	**239.**	**240.**
d	d	a	b	a	a	d	a	c	b
241.	**242.**	**243.**	**244.**	**245.**	**246.**	**247.**	**248.**	**249.**	**250.**
c	a	d	c	d	a	b	c	d	a
251.	**252.**	**253.**	**254.**	**255.**	**256.**	**257.**	**258.**	**259.**	**260.**
a	c	b	a	c	d	a	b	b	c
261.	**262.**	**263.**	**264.**	**265.**	**266.**	**267.**	**268.**	**269.**	**270.**
a	c	b	c	a	d	a	d	a	b